PLANT IDENTIFICATION TERMINOLOGY
AN ILLUSTRATED GLOSSARY

PLANT IDENTIFICATION TERMINOLOGY

AN ILLUSTRATED GLOSSARY

JAMES G. HARRIS
MELINDA WOOLF HARRIS

SECOND EDITION

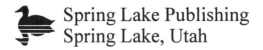
Spring Lake Publishing
Spring Lake, Utah

First Printing, 1st Edition 1994
Seventh Printing, 2nd Edition 2000
Twentieth Printing, 2nd Edition 2022

Publisher's Cataloging in Publication Data

Harris, James G., 1954-
Harris, Melinda Woolf, 1953-
    Plant identification terminology: an illustrated glossary / James G. Harris & Melinda Woolf Harris —
    2nd ed.
    216 p.: illus.; 26 cm.
    ISBN 0-9640221-7-6 (cloth: alk. paper)
    ISBN 0-9640221-6-8 (paperback: alk. paper)
        1. Botany—Dictionaries. 2. Botany—Terminology. I. Title.
QK9.H37 2001   580.3 H242   00-191714

PREFACE TO THE FIRST EDITION

A formidable task facing the student of plant taxonomy is gaining a working knowledge of the vast terminology required to use a typical plant identification key. Most keys are provided with a glossary, but, because of the technical nature of many botanical terms, these glossaries are often of limited value. Either the user may find a verbal description inadequate to convey the essence of a complex botanical term, or the definition may include two or three additional terms that also must be defined to make sense of the original definition. The experience of keying out even one plant specimen may become so tedious and frustrating that the student quickly loses all enthusiasm for plant identification.

Often all that is required to quickly convey the meaning of a term is a simple illustration. In this volume we have attempted to assemble a glossary that includes most of the botanical terms a student would encounter in a typical plant identification key, and we have provided line drawings for all terms that we feel might be made clearer by an illustration.

For simplicity of use, we have attempted, whenever possible, to place illustrations on the same page (or on the facing page) as the term definition. Naturally, this has meant much duplication of some illustrations. For example, the terms "receptacle," "calyx," "corolla," "androecium" and "gynoecium" all could be illustrated with the same drawing placed in a single location in the text. Instead, we have placed copies of the drawing throughout the text near each appropriate definition. We believe that this approach will make the book more convenient and useful.

As we have examined botanical keys and descriptions over the years, we have noticed that the same term is often interpreted quite differently by various authors. For example, the word "pubescent" is often used to refer to hairiness of any kind, yet many botanists prefer to reserve the term for instances of short, downy hairiness.

Similarly, the word "scorpioid" is applied in different ways by different botanists. Some use the term to describe a one-sided cymose inflorescence coiled like the tail of a scorpion, while others use it to describe an inflorescence with a zigzag rachis. Historically, "scorpioid" appears in botanical literature in both connotations for at least the last 150 years.

While current usage seems to favor the zigzag interpretation of "scorpioid," it is difficult to argue with those who choose to apply the term to coiled inflorescences. The original Greek word means, literally, scorpion-like, and a coiled inflorescence is certainly more evocative of a scorpion's tail than is a zigzag inflorescence.

Plant systematics, perhaps more than any other area of botany, includes a strong historical element. It is not uncommon, for example, to hear a taxonomist trace his or her professional roots back to Asa Gray or another prominent botanist of the last century. Perhaps the divergent current usages of some botanical terms have their origins in separate professional clans and lineages, having been passed from teacher to student over several generations.

We have not attempted to resolve these conflicts in interpretation; that is not the purpose of this volume. Instead, we have tried to include divergent usages of terms so that the student of botany can make use of the book no matter the interpretation employed by the author of the identification key or description in use.

The book is divided into two parts. Part One is the essential core of the book. It is an alphabetical glossary of more than twenty-four hundred terms commonly used in plant description and identification. We hope that it will prove useful to professional botanists and students of botany alike. Part Two is designed primarily for the student. Here we have grouped related terms together to facilitate study and comparison.

James G. Harris
Melinda Woolf Harris

PREFACE TO THE SECOND EDITION

While this new edition includes approximately three hundred definitions not found in the first edition, the change most apparent to those familiar with the first edition will be the many new illustrations. Many of these illustrate terms new to this edition, but the majority are simply replacements for old illustrations. In fact, a significant percentage of the illustrations in this edition are new. We believe that these new illustrations not only more accurately depict the terms they represent, but are more visually pleasing as well.

The illustrations in this edition are also more thoroughly labeled than those in the old edition. Again, this should make the illustrations more useful to the reader.

Because we see the illustrations as the real strength of this work, and because we want the book to be useful to readers of all levels, we have liberally illustrated it. We are aware that for many readers the book is over-illustrated — perhaps grossly so for professional botanists — but we decided that for the benefit of students and lay readers, it was better to error on the side of overkill.

Another change that will be apparent to former users of the book is the new name for Part Two. We felt that "Terminology by Category" was more descriptive than the former "Specific Terminology." Other than this nomenclatural innovation, and the substitution of several illustrations, Part Two remains essentially unchanged.

We have been pleased and, we admit, somewhat surprised by the overwhelming reception from botanists for the first edition of *Plant Identification Terminology*. Evidently the book fills a niche. We hope that the second edition will prove to be an even more effective tool.

James G. Harris
Melinda Woolf Harris

CONTENTS

PART ONE

ILLUSTRATED GLOSSARY

ILLUSTRATED GLOSSARY

A- (prefix). Meaning without or lacking.

Abaxial. The side away from the axis. Figure 1. (compare **adaxial**)

Aberrant. Different from the usual; atypical; abnormal.

Abortion. The failure of a structure or organ to develop.

Abortive. Not fully or properly developed; rudimentary. Figure 2.

Figure 1   **Figure 2**

Abrupt. Terminating suddenly. Figure 3. (compare **truncate**)

Abruptly pinnate. Pinnate without an odd leaflet at the tip. Figure 4. (same as **even pinnate**)

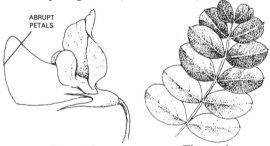

Figure 3   **Figure 4**

Abscission. The falling away of a leaf or other organ caused by the breakdown of thin-walled cells at the base of the structure.

Acantha. A thorn or spine. Figure 5.

Acarpic. Without fruit.

Acarpous. Without carpels; lacking a gynoecium. Figure 6.

Acaulescent. Without a stem, or the stem so short that the leaves are apparently all basal, as in the dandelion. Note: the peduncle should not be confused with the stem. Figure 7.

Accessory bud. An extra bud in a leaf axil. Figure 8.

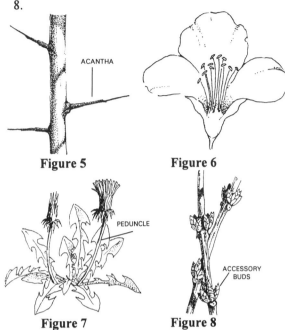

Figure 5   **Figure 6**

Figure 7   **Figure 8**

Accessory fruit. A fleshy fruit developing from a succulent receptacle rather than the pistil. The ripened ovaries are small achenes on the surface of the receptacle, as in the strawberry. Figure 9.

Accrescent. Becoming larger with age, as a calyx which continues to enlarge after anthesis. Figure 10.

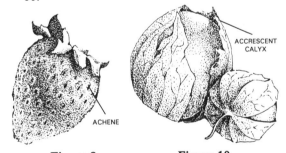

Figure 9   **Figure 10**

Accumbent cotyledons. Cotyledons lying against the radicle along one edge. Figure 11. (compare **incumbent cotyledons**)

Acephalous. Headless, or with the heads much reduced.

3

Aceriform. Shaped like a maple leaf. Figure 12.

Acerose. Needle-shaped, as the leaves of pine or spruce. Figure 13.

Achene. A small, dry, indehiscent fruit with a single locule and a single seed (ovule), and with the seed attached to the ovary wall at a single point, as in the sunflower. Figure 14.

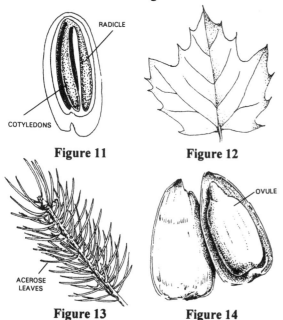

Figure 11   Figure 12

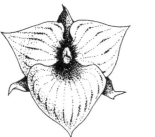

Figure 13   Figure 14

Achenodium. See **schizocarp**.

Achilary. Lacking a lip, as in some orchid blossoms. Figure 15.

Achlamydeous. Lacking a perianth. Figure 16.

Figure 15   Figure 16

Achlorophyllous. Without chlorophyll, as in plants or plant structures which are not green.

Acicula. A small needle-like or bristle-like structure. Figure 17.

Acicular. Needle-shaped. (see **acerose**)

Aciculate. Marked as with pinpricks or needle

scratches; needle-shaped. Figure 13.

Acidophilous. Acid loving, as in a plant which prefers acidic soils.

Acies. The edge of some angled stems. Figure 18.

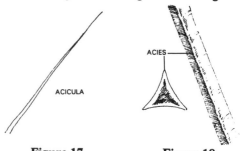

Figure 17   Figure 18

Acorn. The hard, dry, indehiscent fruit of oaks, with a single, large seed and a cuplike base. Figure 19.

Acotyledonous. Without cotyledons.

Acrid. A sharp, bitter, or biting taste.

Acrogen. A non-flowering plant which grows only at the apex, as in a fern.

Acrogenic. See **acrogenous**.

Acrogenous. Growing from the apex.

Acropetal. Near the tip rather than the base; produced sequentially from the base to the apex, as the flowers in an indeterminate inflorescence. Figure 20.

Figure 19   Figure 20

Acroscopic. Facing the tip or apex. Figure 21.

Actinomorphic. Radially symmetrical, so that a line drawn through the middle of the structure along any plane will produce

Figure 21

a mirror image on either side.   Figure 22. (compare **zygomorphic**, and see **regular**)

Actinomorphous. See **actinomorphic**.

Aculeate. Prickly; covered with prickles. Figure 23.

Aculeolate. Minutely prickly; covered with tiny prickles. Figure 24.

Acumen. Apex. Figure 25.

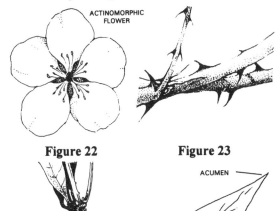

Figure 22   **Figure 23**

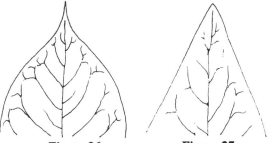

Figure 24   **Figure 25**

Acuminate. Gradually tapering to a sharp point and forming concave sides along the tip. Figure 26.

Acute. Tapering to a pointed apex with more or less straight sides. Figure 27.

Figure 26   **Figure 27**

Acyclic. With the floral parts arranged spirally rather than in whorls. Figure 28.

Ad- (prefix). Meaning to or toward.

Adaxial. The side toward the axis. Figure 29. (compare **abaxial**)

Adenophorous. Gland-bearing.

Adherent. Sticking together of unlike parts, as the anthers to the style. The attachment is not as firm or solid as **adnate**.

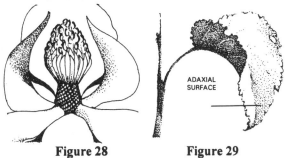

Figure 28   **Figure 29**

Adnate. Fusion of unlike parts, as the stamens to the corolla. Figure 30. (compare **connate**)

Adpressed. Lying close to another organ, but not fused to it.

Adscendent. See **ascending**.

Adsurgent. See **ascending**.

Aduncate. Hooked. Figure 31.

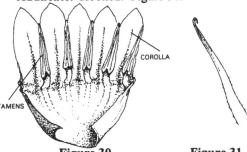

Figure 30   **Figure 31**

Adventitious. Structures or organs developing in an unusual position, as roots originating on the stem. Figure 32.

Adventive. Not native; introduced and beginning to spread in the new region.

Aequilateral. Equal-sided, as opposed to oblique (in leaves). Figure 33.

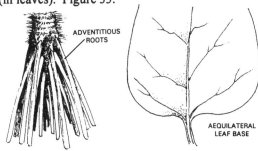

Figure 32   **Figure 33**

Aerial. Occurring above ground or water.

Aestival. Flowering or appearing in the summer.

Aestivation. The arrangement of floral parts in a bud.

Afoliate. Without leaves.

Agamospecies. A species which usually produces seeds asexually, by agamospermy,

Agamospermy. Formation of seed without fertilization.

Agglomerate. Crowded into a dense cluster. Figure 34.

Aggregate. Densely clustered. Figure 34.

Aggregate fruit. Usually applied to a cluster or group of small fleshy fruits originating from a number of separate pistils in a single flower, as in the clustered drupelets of the raspberry. Figure 35.

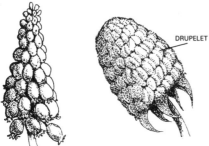

Figure 34   Figure 35

Agynic. Lacking a gynoecium. Figure 6; free from the pistil. Figure 36.

Aianthous. Flowering constantly.

Akene. See achene.

Ala (pl. **alae**). A winglike extension or process; one of the two lateral petals of a papilionaceous corolla. Figure 37.

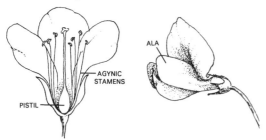

Figure 36   Figure 37

Alate. Winged. Figure 38.

Albumen. The nutritive tissue in a seed. Figure 39.

Albuminous. With albumen. Figure 39.

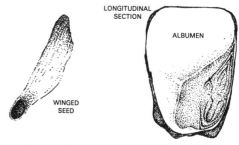

Figure 38   Figure 39

Aliferous. With wings. Figure 38.

Alkaline. Material that is basic rather than acidic; having a Ph greater than 7.0.

Alliaceous. Having the smell or taste of garlic.

Allogamous. Reproducing by cross-fertilization.

Allogamy. Cross-pollination.

Allopatric. Occupying different geographic regions. (compare **sympatric**)

Allotropous flower. A flower shaped so that its nectar is readily available to pollinators.

Alluvial. Of or pertaining to alluvium (i.e. organic or inorganic materials deposited by running water).

Alpine. Of or pertaining to areas above timberline; growing above timberline.

Alternate. Borne singly at each node, as leaves on a stem. Figure 40; borne between rather than over other organs, as stamens between the petals. Figure 41. (compare **opposite**)

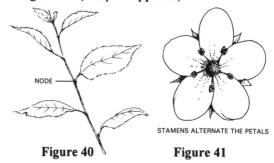

Figure 40   Figure 41

Alveola (pl. **alveolae, alveolas**). Pits arranged in a honeycomb-like pattern. Figure 42.

Alveolar. See **alveolate**.

Alveolate. Honeycombed, with pits separated by thin, ridged partitions. Figure 42.

Alveole. See **alveola**.

Alveolus (pl. **alveoli**). See **alveola**.

Ament. An inflorescence consisting of a dense spike

or raceme of apetalous, unisexual flowers, as in Salicaceae and Betulaceae; a catkin. Figure 43.

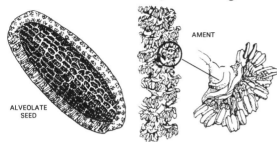

Figure 42   **Figure 43**

Amentaceous. Ament-like, or ament-bearing.

Amentiferous. Ament-bearing.

Amentum. See **ament**.

Amethystine. Amethyst-colored; purplish.

Ammophilous. Sand-loving.

Amorphous. Without any definite form; shapeless; lacking symmetry.

Amphibious. Living both in water and on land.

Amphicarpous. Producing two types of fruit.

Amphitropous ovule. An ovule which is half-inverted and straight, with the hilum lateral. Figure 44.

Amphora. The lower portion of a pyxis. Figure 45.

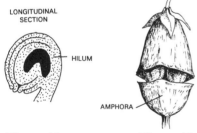

Figure 44   **Figure 45**

Amplectant. See **amplexicaul**.

Amplexicaul. Clasping the stem, as the base or stipules of some leaves. Figure 46.

Ampliate. Enlarged or expanded. Figure 47.

Ampulla. A bladder or swelling. Figure 48.

Ampullaceous. Swelling out like a bottle or bladder. Figure 48.

Anandrous. Without stamens; lacking an androecium. Figure 49.

Ananthous. Without flowers.

Anastomosing. Rejoining after branching and forming an intertwining network, as in some leaf veins. Figure 50.

Anatropous ovule. An ovule which is inverted and straight with the micropyle situated next to the funiculus. Figure 51.

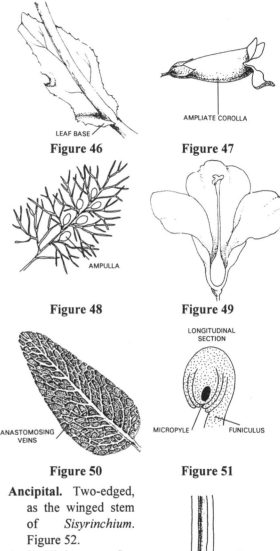

Figure 46   **Figure 47**

Figure 48   **Figure 49**

Figure 50   **Figure 51**

Ancipital. Two-edged, as the winged stem of *Sisyrinchium*. Figure 52.

Androclinium. See **clinandrium**.

Androecium. All of the stamens in a flower, collectively.

Androgynophore. Stalk supporting the androecium and gynoecium in some flowers. Figure 53.

Figure 52

Androgynous. With both staminate and pistilate flowers, the staminate flowers borne above the pistilate, as in the inflorescence of some *Carex* species. Figure 54. (compare **gynaecandrous**)

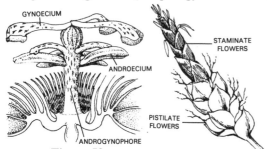

Figure 53   **Figure 54**

Androphore. Stalk supporting a group of stamens. Figure 55.

Andro-dioecious. Having staminate and perfect flowers on separate plants.

Andro-monoecious. See **andro-polygamous**.

Andro-polygamous. Having staminate and perfect flowers on the same plant.

Androspore. A male spore of *Isoetes*. Figure 56.

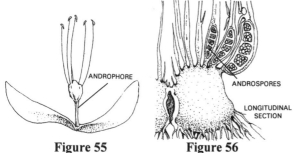

Figure 55   **Figure 56**

Anemophilous. Wind pollinated; producing wind-borne pollen.

Angiosperm. A plant producing flowers and bearing ovules (seeds) in an ovary (fruit).

Angulate. Angled. Figure 57.

Angustiseptate. Of a fruit flattened at right angles to the septum; the septum crosses the narrowest diameter. Figure 58.

Anisomerous. With a different number of parts (usually less) than the other floral whorls, as in a flower with five sepals and petals, but only two stamens. Figure 59.

Annotinal. Appearing annually.

Annual. A plant which germinates from seed, flowers, sets seed, and dies in the same year.

Annular. In the form of a ring. Figure 60.

Annulate. In the form of a ring. Figure 60; with rings or ringlike markings.

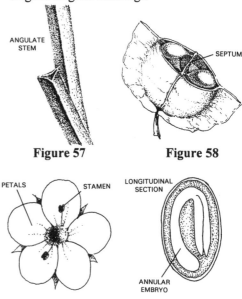

Figure 57   **Figure 58**

Figure 59   **Figure 60**

Annulus. A row of specialized, thick-walled cells along one side of a fern sporangium which aids in the dispersal of spores. Figure 61; a ring-shaped structure.

Anomocytic stomate. A stomate lacking differentiated subsidiary cells.

Antarctic. Distributed in those regions of the earth lying between the Antarctic Circle and the South Pole.

Antepetalous. Directly in front of (opposite) the petals. Figure 62.

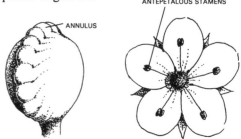

Figure 61   **Figure 62**

Anterior. In the front; on the side away from the axis, as the lower lip of a bilabiate corolla. Figure 63. (compare **posterior**)

Antesepalous. Directly in front of (opposite) the sepals. Figure 64.

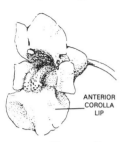

Figure 63

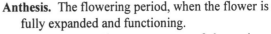

ANTESEPALOUS STAMENS

Figure 64

Anthela. An inflorescence with lateral flowering branches exceeding the main axis, as in some species in the genus *Juncus*. Figure 65.

Anthelate. With the inflorescence in the form of an anthela. Figure 65.

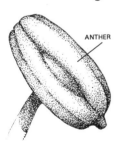

Figure 65

Anthemia. See **anthemy**.

Anthemy. A flower-cluster.

Anther. The expanded, apical, pollen bearing portion of the stamen. Figure 66.

Anther sac. One of the pollen bearing chambers of the anther. Figure 67.

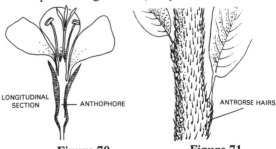

Figure 66   Figure 67

Antherid. See **antheridium**.

Antheridium (pl. **antheridia**). The male reproductive structure in moss and fern gameto-phytes. Figure 68.

Antheriferous. Bearing anthers.

Antheroid. Anther-like.

Antherozoid. Male sexual cells.

Anthesis. The flowering period, when the flower is fully expanded and functioning.

Anthesmotaxis. The arrangement of the various flower parts.

Anthocarp. A fruit with some portion of the flower besides the pericarp persisting, as in a pome with the fleshy perianth tube surrounding the pericarp. Figure 69.

Anthocarpous. Of or pertaining to anthocarps; bearing anthocarps. Figure 69.

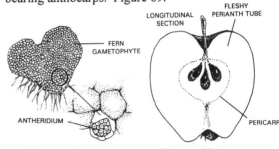

Figure 68   Figure 69

Anthocyanic. Containing anthocyanin pigments.

Anthocyanin. Water-soluble pigments (blue, purple, or red).

Anthophore. An elongated stalk (stipe) bearing the corolla, stamens, and pistil above the receptacle and calyx. Figure 70.

Anthotaxy. The arrangement of flowers on the flowering axis; inflorescence.

Anthoxanthin. Water-soluble pigments (yellow, orange, or red).

Anthracine. Coal-black.

Antipetalous. See **Antepetalous**.

Antisepalous. See **Antesepalous**.

Antrorse (adv. **antrorsely**). Directed forward or upward. Figure 71. (compare **retrorse**)

Figure 70   Figure 71

Aperturate. With one or more openings or aper-tures. In pollen grains, these apertures may be only thin spots rather than actual perforations. Figure 72.

Apetalous. Without petals. Figure 73.

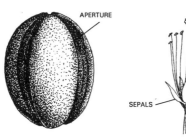

Figure 72   **Figure 73**

Apex (pl. **apices**). The tip; the point farthest from the point of attachment. Figure 74.

Aphyllopodic. Having the lowermost leaves reduced to small scales. Figure 75. (compare **phyllopodic**)

Aphyllous. Without leaves.

Apical. Located at the apex or tip. Figure 76.

Apicula. See **apiculus**.

Apiculate. Ending abruptly in a small, slender point. Figure 77.

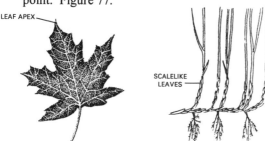

Figure 74   **Figure 75**

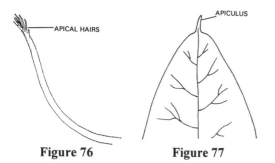

Figure 76   **Figure 77**

Apiculation. See **apiculus**.

Apiculus. A small, slender point. Figure 77.

Apocarp. A flower with carpels forming separate pistils, as in a buttercup. Figure 78. (compare **syncarp**)

Apocarpous. Of or pertaining to apocarps; with separate carpels. Figure 78. (compare **syncarpous**)

Apogamous. See **apomictic**.

Apogamy. See **apomixis**.

Apomictic. Of or pertaining to apomixis.

Apomixis. Defined broadly as any form of asexual reproduction and narrowly, and more commonly, as seed production without fertilization (agamospermy).

Apopetalous. Having separate petals. Figure 79. (same as **polypetalous**; compare **sympetalous** and **gamopetalous**)

Apophysis. A projection or protuberance; that portion of a cone scale that is exposed when the cone is closed. Figure 80.

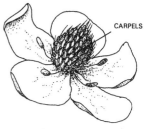

Figure 78

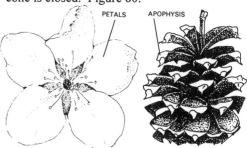

Figure 79   **Figure 80**

Apospory. Development of gametophytes from somatic cells.

Apostemonous. With separate stamens. Figure 81.

Appendage. A secondary part attached to a main structure. Figure 82.

Appendiculate. Bearing appendages. Figure 82.

Applanate. Flattened. Figure 83.

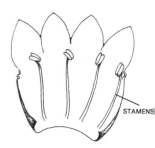

Figure 81

Appressed. Pressed close or flat against another organ. Figure 84.

Approximate. Borne close together, but not fused.

Apterous. Wingless.

Apyrene. Seedless.

Aquatic. Growing in water.

Arachnoid. Bearing long, cobwebby, entangled hairs. Figure 85.

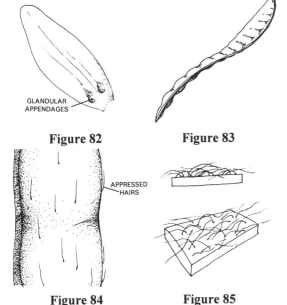

Figure 82

Figure 83

Figure 84

Figure 85

Araneose. See **arachnoid**.

Arboreal. See **arborescent**.

Arboreous. See **arborescent**.

Arborescent. Tree-like.

Arbuscula. A shrub with a tree-like form.

Archegone. See **archegonium**.

Archegonium (pl. **archegonia**). The female reproductive structure in moss and fern gametophytes. Figure 86.

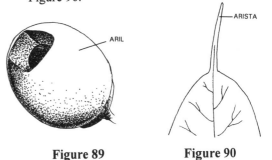

Figure 86

Arctic. Distributed in those regions of the earth lying between the Arctic Circle and the North Pole.

Arcuate. Curved into an arch, like a bow. Figure 87.

Arenaceous. Sandy; growing in sand.

Arenicolous. Growing in sand.

Areola (pl. **areolae, areolas**). A small, well-defined area on a surface, as the area between the veinlets of a leaf or the region of a cactus bearing the flowers and/or spines. Figure 88.

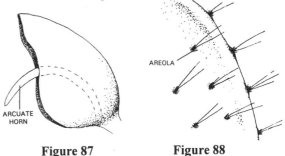

Figure 87

Figure 88

Areole. See **areola**.

Areolate. Marked with areolae. Figure 88.

Argenteous. Silvery.

Argillaceous. Clayey; of or pertaining to plants growing on clay soils.

Argillicolous. Growing on clay soils.

Argute. Sharp.

Arhizous. Without roots.

Aril. An appendage growing at or near the hilum of a seed; fleshy thickening of the seed coat, as in *Taxus*. Figure 89.

Arillate. Possessing an aril. Figure 89.

Arilliform. Aril-like.

Arillode. A false aril.

Arista (pl. **aristae**). An awn or bristle. Figure 90.

Aristate. Bearing an awn or bristle at the tip. Figure 90.

Figure 89

Figure 90

Aristiform. Awn-like.

Aristulate. Bearing a minute awn or bristle at the tip. Figure 91.

Armature. Thorns, spines, barbs, or prickles. Figure 92.

Arm-cell. Cells with incomplete septae extending inward, as in the leaf mesophyll cells of some members of the grass family.

Armed. Bearing thorns, spines, barbs, or prickles. Figure 92.

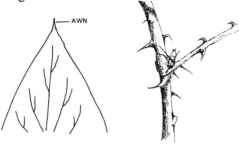

| Figure 91 | Figure 92 |

Article. Section of a fruit separated from others by a constricted joint. Figure 93.

Articulate. Jointed. Figure 93; separating at maturity along a well-defined line of dehiscence. Figure 94.

Articulation. A joint or point of attachment. Figure 93.

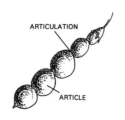

| Figure 93 | Figure 94 |

Arundinaceous. Reed-like in form.

Ascendent. See **ascending**.

Ascending. Growing obliquely upward, usually curved. Figure 95.

Asepalous. Without sepals. Figure 96.

Asexual. Reproducing without sexual union.

Asperity. A tiny projection or hairlike prickle of an epidermal cell. Figure 97.

Asperous. Rough to the touch.

Assumentum (pl. assumenta). A valve of a silique. Figure 98.

Assurgent. See **ascending**.

Astemonous. Without stamens. Figure 99.

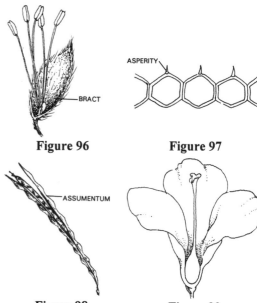

| Figure 96 | Figure 97 |

Figure 95

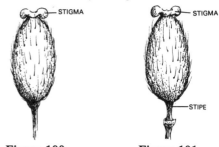

| Figure 98 | Figure 99 |

Astringent. Constricting or contracting.

Astylocarpellous. Lacking a style and a stipe. Figure 100.

Astylocarpepodic. Without a style, but with a stipe. Figure 101.

Astylous. Without a style. Figure 100.

| Figure 100 | Figure 101 |

Asymmetric. Not divisible into equal halves, as in some leaves. Figure 102; irregular in shape.

Atomate. Bearing sessile or subsessile glands. Figure 103.

Atomiferous. See **atomate**.

Atratous. Blackened or turning black.

Atro- (prefix). Dark or blackish.

Atropous ovule. See **orthotropous ovule.**

Atropurpurea. Dark purple, often almost blackish.

Figure 102   **Figure 103**

Attenuate. Tapering gradually to a narrow tip or base. Figure 104.

Atypical. Not typical.

Auricle. A small, ear-shaped appendage. Figure 105.

Auriculate. With auricles. Figure 105.

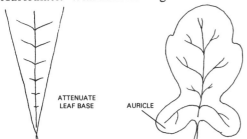

Figure 104   **Figure 105**

Auriculate-clasping. With earlike lobes at the base of a leaf, encircling the stem. Figure 106.

Austral. Southern. (compare **boreal**)

Autocarp. A fruit produced through self-fertilization.

Autogamous. Self-fertilized.

Autogamy. Self-fertilization.

Autophilous. Self-pollinated.

Autophytic. See **autotrophic.**

Autotrophic. Producing its own nutritive substances; containing chlorophyll and, therefore,

Figure 106

green; photosynthetic.

Autumnal. Flowering or appearing in the autumn.

Awl-shaped. Short, narrowly triangular, and sharply pointed like an awl. Figure 107.

Awn. A narrow, bristlelike appendage, usually at the tip or dorsal surface. Figure 108.

Awned. Possessing an awn. Figure 108.

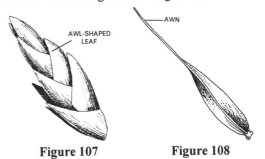

Figure 107   **Figure 108**

Axial. See **axile.**

Axil. The point of the upper angle formed between the axis of a stem and any part (usually a leaf) arising from it. Figure 109.

Axile. Positioned on the axis; pertaining to the axis.

Axile placentation. Ovules attached to the central axis of an ovary with two or more locules. Figure 110.

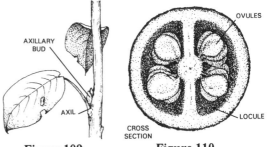

Figure 109   **Figure 110**

Axillary. Positioned in or arising in an axil. Figure 109.

Axis (pl. **axes**). The longitudinal, central supporting structure or line around which various organs are borne, as a stem bearing leaves. Figure 111.

Baccate. Berrylike and soft.

Balausta. A pomegranate fruit. Figure 112

Balsam. A fragrant, sticky exudate from any of various tree species, especially those of the genus *Commiphora.*

Balsamiferous. Producing balsam; balsam-like.

Banded. Striped. Figure 113.

Banner. The upper and usually largest petal of a papilionaceous flower, as in peas and sweet peas. Figure 114.

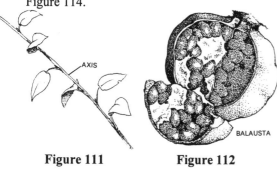

Figure 111   Figure 112

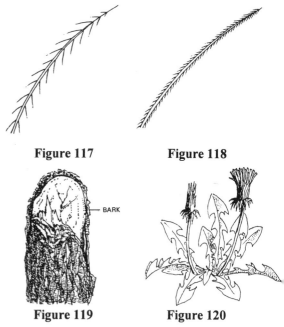

Figure 117   Figure 118

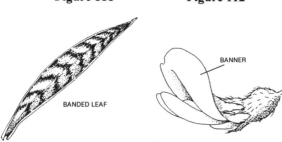

Figure 113   Figure 114

Barbate. Bearded or tufted with long, stiff hairs. Figure 115.

Barbed. With short, rigid, reflexed points, like the barb of a fishhook. Figure 116.

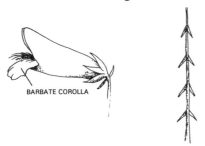

Figure 115   Figure 116

Barbellate. With short, stiff hairs or barbs. Figure 117.

Barbellulate. With very tiny short, stiff hairs or barbs. Figure 118.

Bark. The outermost layers of a woody stem including all of the living and nonliving tissues external to the cambium. Figure 119.

Basal. Positioned at or arising from the base, as leaves arising from the base of the stem. Figure 120.

Figure 119   Figure 120

Basal placentation. Ovules positioned at the base of a single-loculed ovary. Figure 121.

Basifixed. Attached by the base. Figure 122. (compare **versatile** and **dorsifixed**)

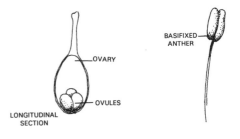

Figure 121   Figure 122

Basinerved. With veins arising from the base. Figure 123.

Basipetal. Near the base rather than the tip; produced sequentially from the apex toward the base, as the flowers in a determinate inflorescence. Figure 124.

Figure 123

Basiscopic. Facing toward the base. Figure 125.

Bast. The fibrous inner bark of some trees; phloem.

Beak. A narrow or prolonged tip, as on some fruits and seeds. Figure 126.

Beaked. Bearing a beak. Figure 126.

Bearded. Bearing one or more tufts of long hairs. Figure 127.

Betaxanthin. Betalain pigment varying from yellow to red.

Bi- (prefix). Meaning two or twice.

Bicarpellary. See **bicarpellate**.

Bicarpellate. With two carpels. Figure 130.

Bicolored. Of two distinct colors.

Biconcave. Concave on both sides. Figure 131.

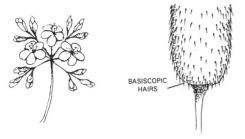

Figure 124   **Figure 125**

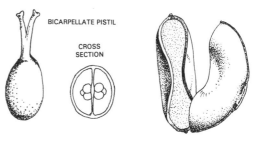

Figure 130   **Figure 131**

Biconvex. Convex on both sides. Figure 132.

Bicrenate. Doubly crenate, as when the teeth of a crenate leaf are also crenate. Figure 133.

Bicuspidate. With two sharp points. Figure 134.

Bidentate. With two teeth. Figure 135.

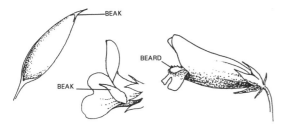

Figure 126   **Figure 127**

Belly (or **bellying**). A swelling on one side, as in some corollas in the Labiatae. Figure 128.

Berry. A fleshy fruit developing from a single pistil, with several or many seeds, as the tomato. Sometimes applied to any fruit which is fleshy or pulpy throughout, i.e. lacking a pit or core. Figure 129.

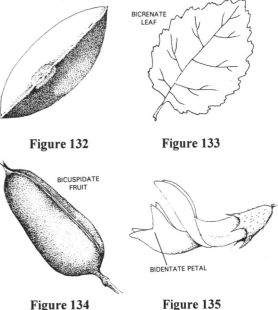

Figure 132   **Figure 133**

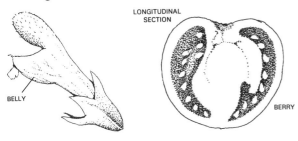

Figure 128   **Figure 129**

Betacyanin. Betalain pigment varying from blue to red.

Betalain. Water-soluble, nitrogen-containing pigments.

Figure 134   **Figure 135**

Biduous. Lasting two days.

Biennial. A plant which lives two years, usually forming a basal rosette of leaves the first year and flowers and fruits the second year.

Bifacial. With the opposite surfaces different in color or texture, as in some leaves.

Bifarious. In two vertical rows. Figure 136.

Biferous. Appearing twice annually.

Bifid. Deeply two-cleft or two-lobed, usually from the tip. Figure 137.

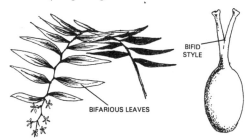

Figure 136   **Figure 137**

Biflorous. Flowering in the spring and again in the autumn.

Bifoliate. With two leaves or two leaflets. Figure 138.

Bifurcate. Two-forked; divided into two branches. Figure 139.

Bigeminate. Twice divided into equal pairs. Figure 140.

Bijugate. See **bigeminate**.

Bijugous. See **bigeminate**.

Bilabiate. Two-lipped, as in many irregular flowers. Figure 141.

Bilateral. Arranged on two sides, as leaves on a stem. Figure 142.

Bilobate. See **bilobed**.

Bilobed. Divided into two lobes. Figure 143.

Bilocellate. Divided into two locelli or secondary locules, as when a main locule of an ovary is partitioned into two cavities. Figure 144.

Bilocular. With two locules, as in some ovaries. Figure 145.

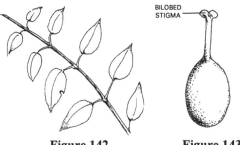

Figure 142   **Figure 143**

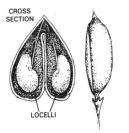

Figure 144   **Figure 145**

Biloculate. See **bilocular**.

Bimestrial. Lasting two months; occurring every two months.

Binate. Borne in pairs. Figure 146.

Bipalmate. Twice palmate; with the divisions again palmately divided. Figure 147.

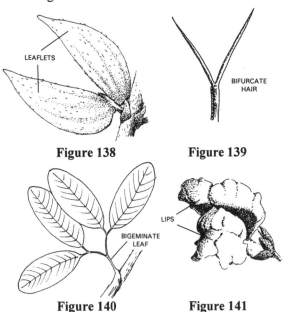

Figure 138   **Figure 139**

Figure 140   **Figure 141**

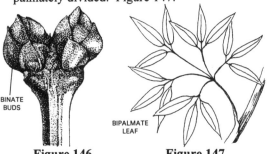

Figure 146   **Figure 147**

Bipartite. Divided almost to the base into two divisions. Figure 148.

Bipetalous. With two petals.

Bipinnate. Twice pinnate; with the divisions again pinnately divided. Figure 149.

Figure 148   **Figure 149**

Bipinnatifid. Twice pinnately cleft. Figure 150.

Bis- (prefix). See **bi-**.

Bisected. Split into two parts. See **bifid** and **bipartite**.

Biserial. See **biseriate**.

Biseriate. Arranged in two rows or series. Figure 151.

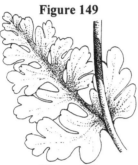

Figure 150

Biserrate. Doubly serrate, as when the teeth of a serrate leaf are also serrate. Figure 152.

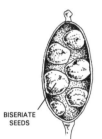

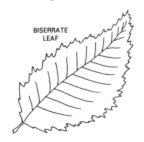

Figure 151   **Figure 152**

Bisexual. A flower with both male and female reproductive organs (stamens and pistils). Figure 153. (same as **perfect**)

Biternate. Doubly ternate with the ternate divisions again ternately divided. Figure 154.

Biturbinate. Top-shaped, but with the widest part some distance from one end. Figure 155.

Bivalvular. With two valves. Figure 156.

Bladder. A structure which is thin-walled and inflated. Figure 157.

Bladderlike. Thin-walled and inflated. Figure 157.

Bladdery. See **bladderlike**.

Blade. The broad part of a leaf or petal. Figure 158.

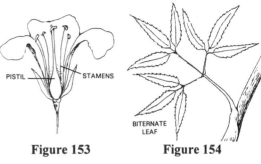

Figure 153   **Figure 154**

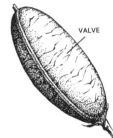

Figure 155   **Figure 156**

Figure 157   **Figure 158**

Bloom. A whitish, waxy, powdery coating on a surface; the flower.

Blossom. A flower.

Blotched. Marked with irregular spots or blots. Figure 159.

Bole. The trunk of a tree. Figure 160.

Bordered. With the edge of a different

Figure 159

color than the main body of the organ or structure. Figure 161.

Boreal. Northern. (compare **austral**)

Boss. A protuberance or projection from a surface or organ. Figure 162.

Botuliform. Sausage-shaped. Figure 163.

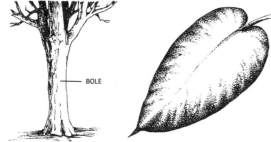

Figure 160   **Figure 161**

Figure 162   **Figure 163**

Brachiate. With paired branches diverging from the stem at nearly right angles. Figure 164.

Brachyblast. A short, spur branch. Figure 165.

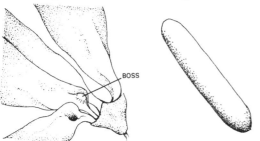

Figure 164   **Figure 165**

Brackish. Somewhat saline.

Bract. A reduced leaf or leaflike structure at the base of a flower or inflorescence. Figure 166; in conifers, one of the main structures arising from the cone axis. Figure 167.

Bracteal. Of or pertaining to bracts; bracteate.

Bracteate. With bracts.

Bracteiform. Bractlike.

Bracteody. With bracts mimicking floral whorls. Figure 168.

Bracteolate. With bracteoles. Figure 169.

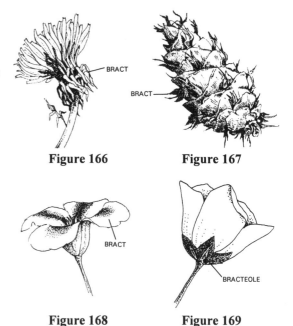

Figure 166   **Figure 167**

Figure 168   **Figure 169**

Bracteole. A small bract, often secondary in nature; a bractlet. Figure 169.

Bracteose. With many bracts or with conspicuous bracts.

Bractlet. See **bracteole**.

Branch. A major division of the stem or trunk. Figure 170.

Branchlet. A small branch growing from a larger branch. Figure 170.

Bridge. A band of tissue connecting the corolla scales, as in *Cuscuta*. Figure 171.

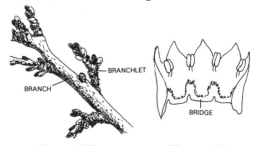

Figure 170   **Figure 171**

Bristle. A short, stiff hair or hairlike structure. Figure 172.

Bristly. Covered with bristles. (same as **setose**)

Brunescent. Brownish.

Bud. An undeveloped shoot or flower. Figure 173.

Bud scales. Modified scale-like leaves covering a

bud. Figure 173.

Bulb. An underground bud with thickened fleshy scales, as in the onion. Figure 174.

Bulbel. See **bulbil**.

Bulbiferous. Producing bulbs.

Bulbil. A small bulb arising from the base of a larger bulb. Figure 175.

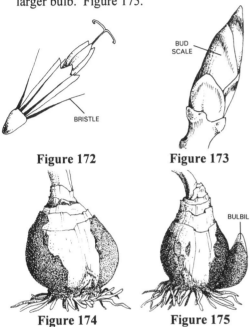

Figure 172

Figure 173

Figure 174

Figure 175

Bulblet. A small bulb; a bulblike structure borne above ground, usually in a leaf axil. Figure 176; bulbil.

Bulbose. Bulblike; with bulbs; of or pertaining to bulbs.

Bulbous. See **bulbose**.

Bulbule. A little bulb.

Bulbus. See **bulbose**.

Figure 176

Bullate. With rounded, blistery projections covering the surface, as in a leaf with the surface raised above the veins. Figure 177.

Bullation. A bullate structure. Figure 177.

Bulliform cells. Large, thin-walled epidermal cells of the intercostal zone of the leaf blade in some members of the grass family. Figure 178.

Bundle scar. Scar left on a twig by the vascular bundles when a leaf falls. Figure 179.

Bur (or **burr**). A structure armed with often hooked or barbed spines or appendages. Figure 180.

Bursicle. A pouch-like or purse-like structure. Figure 181.

Bursiculate. Pouch-like or purse-like in form. Figure 181.

Bursicule (pl. **bursicula**). A small bursicle.

Bursiform. See **bursiculate**.

Bush. See **shrub**.

Buttressed. With props or supports, as in the flared trunks of some trees. Figure 182.

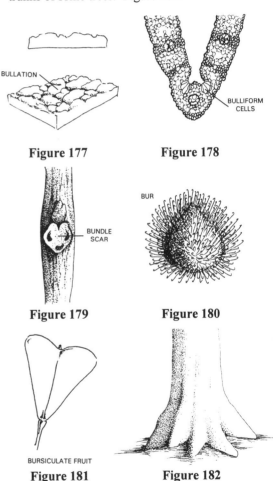

Figure 177

Figure 178

Figure 179

Figure 180

Figure 181

Figure 182

Caducous. Falling off very early compared to similar structures in other plants.

Caerulescent. Bluish.

Caespitose. Growing in dense tufts. Figure 183.

Calathiform. Basket-shaped or cup-shaped. Figure 184.

Calcar. A spur or spurlike appendage. Figure 185.

Calcarate. With a calcar; spurred. Figure 185.

Calceiform. See **calceolate**.

Calceolate. Shoe-shaped or slipper-shaped, as the labellum of some orchids. Figure 186.

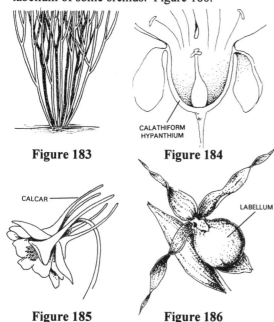

Figure 183   **Figure 184**

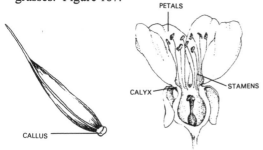

Figure 185   **Figure 186**

Calcicole. A plant growing on calcareous soil.

Calcifuge. A plant which avoids calcareous soil.

Calciphilous. Lime-loving.

Calicate. See **calycate**.

Caliciform. Cup-shaped. Figure 184.

Callose. See **callous**.

Callosity. A hardened or thickened area; the condition of being callous.

Callous. Hardened or thickened; having a callus.

Callus. A hard thickening or protuberance; the thickened basal extension of the lemma in many grasses. Figure 187.

Figure 187   **Figure 188**

Calycanthemous. With a petaloid calyx.

Calycate. With a calyx.

Calycifloral. See **calyciflorous**.

Calyciflorate. See **calyciflorous**.

Calyciflorous. With the petals and stamens adnate to the calyx. Figure 188.

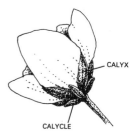

Figure 189

Calyciform. Calyx-like in form.

Calycinal. See **calycine**.

Calycine. Of or pertaining to the calyx; calyx-like.

Calycle. A row of bracts around the calyx, resembling an outer calyx. Figure 189.

Calycoid. Calyx-like.

Calyculate. With small bracts around the calyx, as if possessing an outer calyx. Figure 189; with small bracts around the base of the involucre.

Calycule. See **calycle**.

Calyptra. A hood or lid. Figure 190.

Calyx (pl. **calyces, calyxes**). The outer perianth whorl; collective term for all of the sepals of a flower. Figure 191.

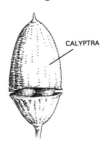

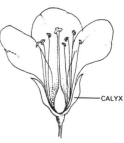

Figure 190   **Figure 191**

Calyx limb. See **calyx lobe**.

Calyx lobe. One of the free portions of a calyx of united sepals. Figure 192.

Calyx tooth. See **calyx lobe**.

Calyx tube. The tube-like united portion of a calyx of united sepals. Figure 192.

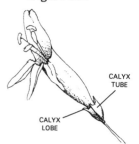

Figure 192

Cambium. A tissue composed of cells capable of active cell division, producing xylem to the inside of the plant and phloem to the outside; a lateral meristem. Figure 193.

Campanulate. Bell-shaped. Figure 194.

Campylotropous ovule. An ovule which is curved so that the micropyle is positioned near the funiculus and the chalaza. Figure 195.

Canaliculate. With longitudinal channels or grooves. Figure 196.

Canescence. A covering of short, fine gray or white hairs producing a gray or white color. Figure 199.

Canescent. Gray or white in color due to a covering of short, fine gray or white hairs. Figure 200.

Figure 199

Capillary. Hair-like; very slender and fine. Figure 201.

Capitate. Head-like, or in a head-shaped cluster, as the flowers in many plant groups (Figure 202), but especially those in the dense capitate inflorescences of the Compositae (Asteraceae) Figure 203.

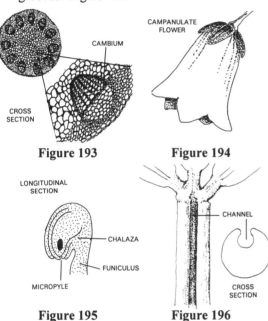

Figure 193   **Figure 194**

Figure 195   **Figure 196**

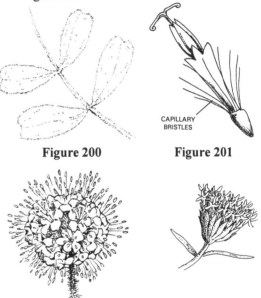

Figure 200   **Figure 201**

Figure 202   **Figure 203**

Cancellate. Latticed with a fine, regular, reticulate pattern. Figure 197.

Cane. A slender, hollow, and often jointed stem, as in a reed. Figure 198; any straight, woody stem arising directly from the ground, as in the raspberry.

Capitellate. With small head-like structures, or with parts in very small head-shaped clusters. Figure 204.

Capitulum. A small flower head. Figure 204.

Capreolate. With tendrils. Figure 205.

Capsular. Of or pertaining to a capsule; capsule-like.

Capsule. A dry, dehiscent fruit composed of more than one carpel. Figure 206.

Figure 197   **Figure 198**

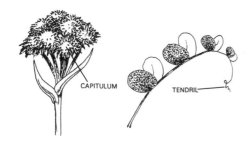

Figure 204

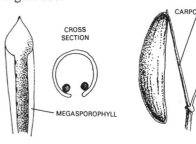

Figure 205

Carina. A keel or ridge. Figures 207 and 208.

Carinal. See **carinate**.

Carinate. Keeled with one or more longi- tudinal ridges. Fig- ures 207 and 208.

Cariopsis. See **cary- opsis**.

Carneous. Flesh-color.

Figure 206

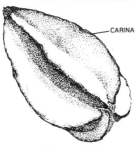

Figure 207

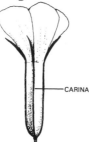

Figure 208

Carnose. With a fleshy texture.

Carotene. Yellow, orange, or red fat-soluble pigments.

Carotenoid pigment. A carotene or xanthophyll pigment.

Carpel. A megasporophyll. Figure 209; a simple pistil formed from one modified leaf, or that part of a compound pistil formed from one modified leaf. Carpel number of a compound pistil is determined by counting the number of stigmas, styles, locules, and placentae. Carpel number is indicated by whichever of these parts is found in the greatest number.

Carpellate. Of or pertaining to carpels; with carpels.

Carpophore. A slender prolongation of the receptacle forming a central axis between the carpels, as in the fruits of some members of the Umbelliferae (Apiaceae) and the Geraniaceae. Figure 210.

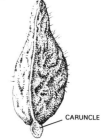

Figure 209   **Figure 210**

Carpopodium. A stipe supporting an ovary. Figure 211. (same as **gynophore**)

Cartilaginous. Tough and firm but elastic and flexible, like cartilage.

Caruncle. A protuberance or appendage near the hilum of a seed. Figure 212.

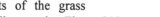

Figure 211   **Figure 212**

Caruncular. Of or per- taining to a caruncle; caruncle-like.

Carunculate. With car- uncles.

Caryopsis. A dry, one- seeded, indehiscent fruit with the seed coat fused to the pericarp, as in the fruits of the grass family; a grain. Figure 213.

Figure 213

Castaneous. Chestnut-colored; dark reddish-brown in color.

Cataphyll. Brown or colorless scale-like structures believed to be modified leaves.

Cataphylloid. Cataphyll-like.

Catkin. An inflorescence consisting of a dense

spike or raceme of apetalous, unisexual flowers as in Salicaceae and Betulaceae; an ament. Figure 214.

Caudate. With a tail-like appendage. Figure 215.

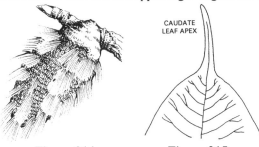

Figure 214   **Figure 215**

Caudex (pl. **caudices, caudexes**). The persistent and often woody base of a herbaceous perennial. Figure 216.

Caudicle. The stalk connecting the pollinia to the stigma in the Orchidaceae. Figure 217.

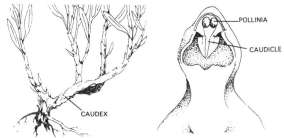

Figure 216   **Figure 217**

Caulescent. With an obvious leafy stem rising above the ground. Figure 218. (Compare **acaulescent**)

Caulicle. A small stem; a rudimentary stem.

Cauliferous. With a stem or stalk.

Cauliflorous. Bearing flowers on the stem or trunk. Figure 219.

Cauliflory. The production of flowers on the stem or trunk. Figure 219.

Cauliform. Stem-like.

Cauline. Of, on, or pertaining to the stem, as leaves arising from the stem above ground level. Figure 220.

Caulis. The main stem of a herbaceous plant.

Caulocarpic. With the stem living for several years.

Cauloid. Stem-like.

Cecidium. An abnormal growth caused by insects or fungal infection. Figure 221.

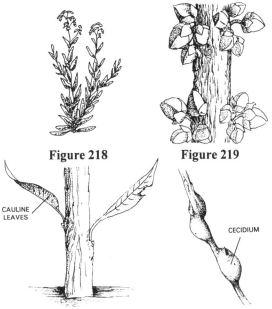

Figure 218   **Figure 219**

Figure 220   **Figure 221**

Cell. As used in plant identification, a hollow cavity or compartment within a structure, as the cavity of the anther containing pollen or the cavity of the ovary containing ovules; a locule. Figure 222.

Cellular. Made up of small cavities or compartments.

Cement Disk. See **retinaculum**.

Cenanthy. The absence of stamens and pistils in a flower, i.e. the perianth is empty.

Centrifugal inflorescence. A flower cluster developing from the center outward, as in a cyme. Figure 223.

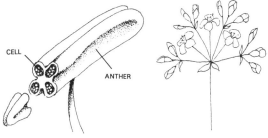

Figure 222   **Figure 223**

Centripetal inflorescence. A flower cluster developing from the edge toward the center, as in a corymb. Figure 224.

Ceraceous. Waxy in texture or appearance.

Ceriferous. See **cerogenous**.
Cernuous. Drooping or nodding. Figure 225.

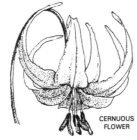
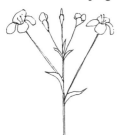

CERNUOUS FLOWER

Figure 224   **Figure 225**

Cerogenous. Wax producing.
Cerulescent. See **caerulescent**.
Cespitose. See **caespitose**.
Chaff. Thin dry scales or bracts, as the bracts on the receptacle of the heads of the Compositae (Asteraceae). Figure 226.
Chaffy. With chaff; chaff-like.
Chalaza. The part of an ovule or seed where the integuments are connected to the nucellus, at the opposite end from the micropyle. Figure 227.

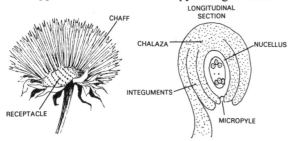

CHAFF

RECEPTACLE

LONGITUDINAL SECTION

CHALAZA   NUCELLUS

INTEGUMENTS

MICROPYLE

Figure 226   **Figure 227**

Chamaephyte. A plant which produces resting buds just above the ground.
Chambered. With hollow spaces. Figure 228.
Channeled. With one or more deep longitudinal grooves. Figure 229.

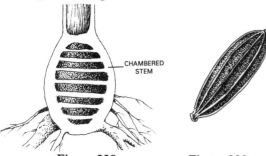

CHAMBERED STEM

CILIA

Figure 228   **Figure 229**

Chaparral. A vegetation type consisting of dense thickets of evergreen shrubs.
Chartaceous. With a papery texture, usually not green.
Chasmogamous. Applied to flowers which open before fertilization and are usually cross-pollinated. (compare **cleistogamous**)
Chasmogamy. The state or condition of being chasmogamous.
Chlamydeous. With, or pertaining to, a floral whorl.
Chloranthous. With green, leaf-like flowers.
Chloranthy. The state or condition of having green, leaf-like flowers.
Chlorophyll. The green pigment of plants associated with photosynthesis.
Chlorophyllous. Of or containing chlorophyll; green.
Chlorotic. Lacking chlorophyll.
Choripetalous. See **apopetalous** or **polypetalous**.
Chorisantherous. See **apostemonous**.
Chorisepalous. See **polysepalous**.
Chrysanthine. See **chrysanthous**.
Chrysanthous. With yellow flowers.
Chrysocarpous. With yellow fruit.
Cicatrice. A scar, such as the scar produced when a leaf separates from the stem. Figure 230.
Cicatrix. See **cicatrice**.
Ciliate. With a marginal fringe of hairs. Figure 231.
Ciliolate. With a marginal fringe of minute hairs. Figure 232.

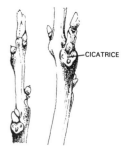

CICATRICE

Figure 230

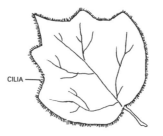

Figure 231   **Figure 232**

Cilium (pl. **cilia**). A small hair or hairlike process, usually along the margin of a structure. Figure 231.

Cincinnus. A dense helicoid cyme with the pedicels short on the developed side. Figures 233 and 234.

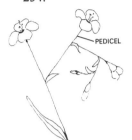
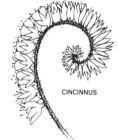

Figure 233   **Figure 234**

Cinereous. Ash-colored; grayish due to a covering of short hairs.

Circinate. Coiled from the tip downward, as in the young leaves of a fern. Figure 235.

Circum- (prefix). Meaning around, as around an object or structure.

Circumscissile. Dehiscing along a transverse circular line, so that the top separates like a lid. Figure 236.

CIRCUMSCISSILE CAPSULE

Figure 235   **Figure 236**

Cirrate. With cirri. Figure 237.

Cirrhiferous. See **cirriferous**.

Cirrhose. See **cirrose**.

Cirriferous. Bearing a tendril. Figure 237.

Cirrose. With cirri. Figure 237; resembling a cirrus. Figure 238.

Cirrus (pl. **cirri**). A

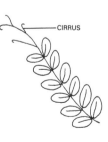

CIRRUS

Figure 237

tendril. Figure 237.

Citreous. Lemon-yellow.

Cladode. See **cladophyll**.

Cladodium (pl. **cladodia**). See **cladophyll**.

Cladophyll. A stem with the form and function of a leaf. Figures 239 and 240. (same as **phylloclade**)

Cladoptosic. Dropping the leaves, branches, and stems at one time, as in *Taxodium*.

Figure 238

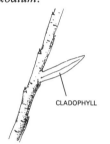

CLADOPHYLL

Figure 239   **Figure 240**

Clambering. Weakly climbing on other plants or surrounding objects.

Clammy-pubescent. With sticky glandular hairs.

Clasping. Wholly or partly surrounding the stem. Figure 241.

Clathrate. Lattice-like in appearance. Figure 242.

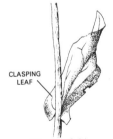

CLATHRATE SEED

CLASPING LEAF

Figure 241   **Figure 242**

Clavate. Club-shaped, gradually widening toward the apex. Figures 243 and 244.

Clavellate. Diminutive of clavate.

Clavicle. A tendril. Figure 237.

Claviculate. With tendrils. Figure 237.

Claviform. See **clavate**.

Clavuncle. A stigmatic cap. Figure 245.

Claw. The narrowed base of some petals and sepals. Figure 246.

Cleft. Cut or split about half-way to the middle or base. Figures 247 and 248.

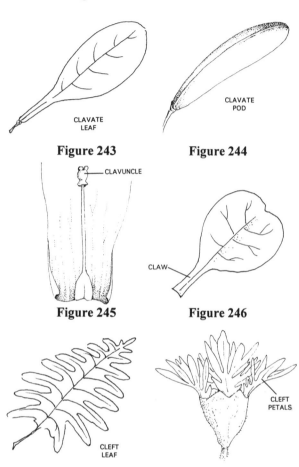

Figure 243   **Figure 244**

Figure 245   **Figure 246**

Figure 247   **Figure 248**

Cleistogamous. Flowers which self-fertilize without opening. (compare **chasmogamous**)

Cleistogamy. The state or condition of being cleistogamous.

Cleistogene. A plant which bears cleistogamous flowers.

Cleistogenous. See **cleistogamous**.

Cleistogeny. See **cleistogamy**.

Climbing. Growing more or less erect by leaning or twining on another structure for support. Figure 249.

Clinandrium. The portion of an orchid column in which the anther is concealed. Figure 250.

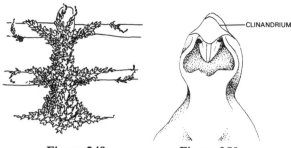

Figure 249   **Figure 250**

Clinanthium. The inflorescence receptacle in the Compositae (Asteraceae). Figure 251.

Clinanthus. See **clinanthium**.

Clinium. See **clinanthium**.

Clone. A group of individuals originating from a single parent plant by vegetative reproduction.

Clouded. Blended with patches of another color.

Coalescence. The state or condition of being coalescent.

Coalescent. United together to form a single unit. Figure 252.

Coarctate. Densely pressed together. Figure 253.

Coat. The covering of a seed. Figure 254; the outer covering of an organ or structure.

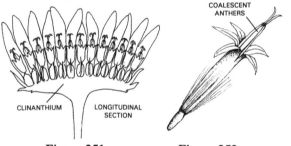

Figure 251   **Figure 252**

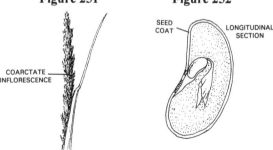

Figure 253   **Figure 254**

Coccus (pl. **cocci**). One of the segments (carpels) of a dry fruit, such as a schizocarp. Figure 255; a berry.

Cochleate. Shaped like the coiled shell of a snail. Figure 256.

Coelospermous. Hollow-seeded; with the seeds or seed-like carpels hollowed on one side. Figure 257.

Coenocarpium. A fruit formed from an entire inflorescence, as in the fig or pineapple; a syconium or multiple fruit. Figure 258.

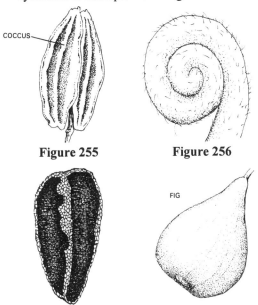

Figure 255   **Figure 256**

Figure 257   **Figure 258**

Coerulean. Blue or bluish.

Coerulescent. See **coerulean**.

Coetaneous. With the leaves and flowers developing at the same time.

Coherent. United by cohesion. Figure 259.

Cohesion. Sticking together of like parts. The attachment is not as firm or solid as **connate**.

Coleoptile. The sheath protecting the stem tip in monocotyledons. Figure 260.

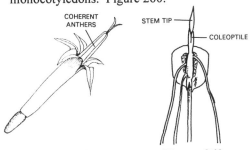

Figure 259   **Figure 260**

Coleorhiza. The sheath which surrounds and is penetrated by the radicle in some seeds. Figure 261.

Collar. The area on the outside of a grass leaf at the juncture of the blade and sheath. Figure 262.

Collateral. Situated side by side. Figure 263.

Colleter. A glandular hair. Figure 264.

Colonial. Forming colonies; usually refers to groups of plants connected to one another by underground organs.

Column. A structure formed by the union of staminal filaments, as in many members of the Malvaceae; the united filaments and style in the Orchidaceae. Figures 265 and 266.

Columnar. Shaped like a column.

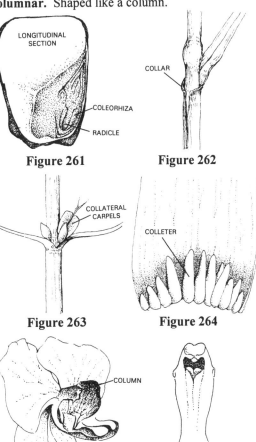

Figure 261   **Figure 262**

Figure 263   **Figure 264**

Figure 265   **Figure 266**

Coma. A tuft of hairs, especially on the tip of a seed. Figure 267; the tuft of bracts on a pineapple. Figure 268; the head of a tree. Figure 269.

Comal. Of or pertaining to a coma.

Comatose. See **comose**.

Commissural. Of or pertaining to a commissure.

Commissure. The face by which two carpels join one another, as in the Umbelliferae (Apiaceae). Figure 270.

Comose. With a coma. Figures 267, 268 and 269; of or pertaining to a coma; coma-like.

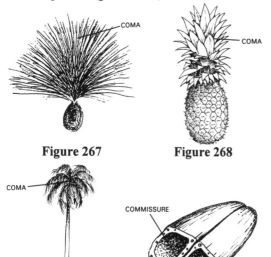

Figure 267   **Figure 268**

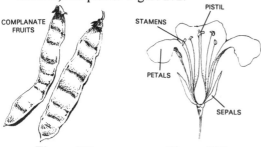

Figure 269   **Figure 270**

Complanate. Flattened. Figure 271.

Complete. With all of the parts typically belonging to it present, as a flower with sepals, petals, stamens, and pistils. Figure 272.

Figure 271   **Figure 272**

Complicate. Folded together. (see **conduplicate**)

Compound. With two or more like parts in one organ.

Compound leaf. A leaf separated into two or more distinct leaflets. Figure 273.

Compound ovary. An ovary of two or more carpels. Figure 274.

Figure 273   **Figure 274**

Compressed. Flattened. Figure 271.

Concave. Hollowed out or curved inward. Figure 275.

Concavo-concave. Concave on both sides. Figure 276.

Concavo-convex. Concave on one side and convex on the other. Figure 277.

Concolored (adj. **concolorous**). With all parts of uniform color.

Conduplicate. Folded together lengthwise with the upper surface within, as the leaves of many grasses. Figure 278.

Cone. A dense cluster of sporophylls on an axis; a strobilus. Figure 279.

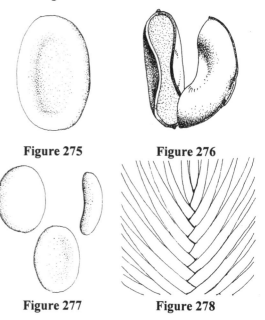

Figure 275   **Figure 276**

Figure 277   **Figure 278**

Conelet. A small cone.

Confert. See **congested**.

Confluent. Running together or blending of one part into another. Figure 280.

Congested. Densely crowded. Figure 281.

Conglomerate. Densely clustered. Figure 281.

Conic. Cone-shaped, with the point of attachment at the broad end. Figure 282.

Connate-perfoliate. With the bases of opposite leaves fused around the stem. Figure 285.

Connective. The portion of the stamen connecting the two pollen sacs of an anther. Figure 286.

Connivent. Converging, but not actually fused or united. Figure 287.

Conocarp. A composite fruit of many carpels on a conical receptacle, as the strawberry. Figure 288.

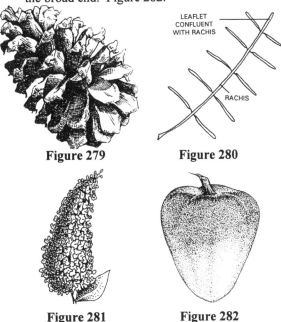

Figure 279

Figure 280

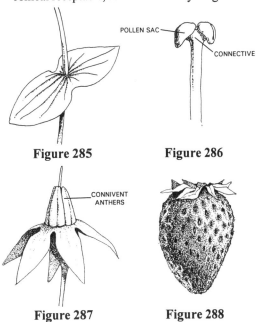

Figure 285

Figure 286

Figure 281

Figure 282

Figure 287

Figure 288

Conical. See **conic**.

Coniferous. Bearing cones or strobili.

Conjugate. Coupled; in a single pair, as in a compound leaf with only two leaflets. Figure 283.

Connate. Fusion of like parts, as the fusion of staminal filaments into a tube. Figure 284. (compare **adnate**)

Conocarpium. See **conocarp**.

Conocarpous. With a conic fruit. Figure 282.

Conoidal. See **conic**.

Conopodium. A conical receptacle. Figure 289.

Consimilar. Similar to one another.

Conspecific. Of the same species.

Constipate. Crowded together. Figure 281.

Constricted. Drawn together or narrowed. Figure 290.

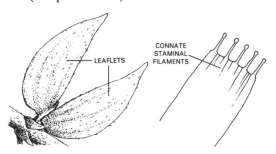

Figure 283

Figure 284

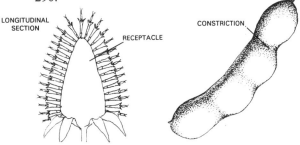

Figure 289

Figure 290

Contiguous. Adjoining; touching.

Continuous. Not jointed; not separating at maturity along a well-defined line of dehiscence.

Contorted. Twisted or bent. Figure 291; convolute.

Contracted. Narrowed; narrow, thick, and dense, as an inflorescence with crowded, short or appressed branches. Figure 292.

Convergent. Meeting together, as leaf veins which come together at the leaf apex. Figure 293.

Convex. Rounded and curved outward on the surface. Figure 294.

Convolute. Rolled up longitudinally. Figure 295; with parts in an overlapping arrangement like shingles on a roof, as petals arranged as to be partially covered by one adjacent petal and partially overlapping the other adjacent petal. Figure 296.

Copious. Large in number or quantity; abundant.

Coppice. A thicket of bushes or small trees; sprouts arising from a stump.

Copse. A thicket of bushes or small trees; woods.

Coracoid. Shaped like the beak of a crow. Figure 297.

Coralloid. Coral-like. Figure 298.

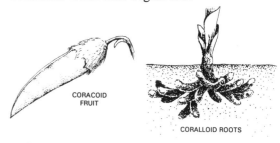

Figure 297   **Figure 298**

Cordate. Heart-shaped, with the notch at the base. Figure 299.

Cordiform. See **cordate**.

Coreaceous. See **coriaceous**.

Coriaceous. With a leathery texture.

Corm. A short, solid, vertical underground stem with thin papery leaves. Figure 300. (compare **bulb**)

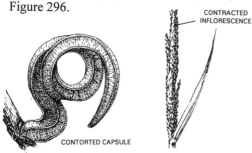

Figure 291   **Figure 292**

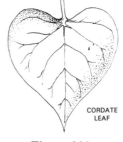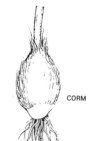

Figure 293   **Figure 294**

Figure 299   **Figure 300**

Cormatose. See **cormous**.

Cormel. A small corm arising at the base of a larger corm.

Cormoid. Corm-like.

Cormous. With corms.

Corneous. Horny. Figure 301.

Cornet. A horn-like structure. Figure 301.

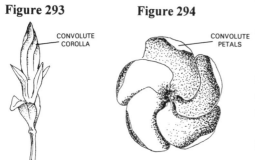

Figure 295   **Figure 296**

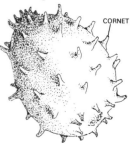

Figure 301

Corniculate. With small horn-like protuberances. Figure 301.

Cornute. Horned. Figure 301.

Corolla. The collective name for all of the petals of a flower; the inner perianth whorl. Figure 302.

Corolla lobe. One of the free portions of a corolla of united petals. Figure 303.

Corolla tube. The hollow, cylindric portion of a corolla of united petals. Figure 303.

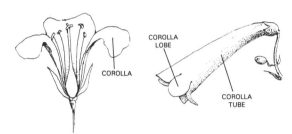

Figure 302   **Figure 303**

Corollate. With a corolla.

Corolliferous. See **corollate**.

Corolliform. See **corolloid**.

Corolloid. Corolla-like in appearance.

Corona. Petal-like or crown-like structures between the petals and stamens in some flowers. Figure 304; a crown. Figure 305.

Coronate. With a corona.

Coroniform. Crown-shaped. Figures 304 and 305.

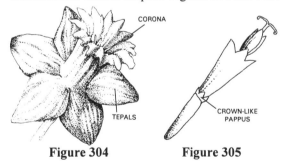

Figure 304   **Figure 305**

Corpusculum. A gland associated with the pollinium in the Asclepiadaceae. Figure 306.

Corrugated. Wrinkled or folded into alternating furrows and ridges. Figures 307 and 308.

Corrugation. A wrinkle, fold, furrow, or ridge. Figures 307 and 308.

Cortex. Bark or rind; root tissue between the epidermis and the stele. Figure 309.

Cortical. Of or pertaining to the cortex.

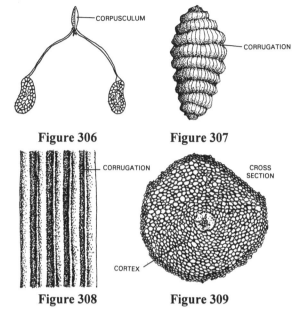

Figure 306   **Figure 307**

Figure 308   **Figure 309**

Corymb. A flat-topped or round-topped inflorescence, racemose, but with the lower pedicels longer than the upper. Figures 310 and 311.

Corymbiform. An inflorescence with the general appearance, but not necessarily the structure, of a true corymb.

Corymbose. Having flowers in corymbs. The term is sometimes used in the same sense as **corymbiform**.

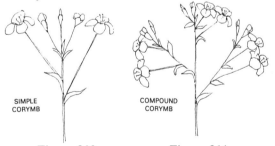

Figure 310   **Figure 311**

Costa (pl. **costae**). A rib or prominent mid-vein. Figure 312.

Costate. Ribbed. Figure 312.

Costular. Pertaining to the ribs or veins.

Cotyledon. A primary leaf of the embryo; a seed leaf. Figures 313 and 314.

Cotyliform. Cup-shaped. Figure 315.

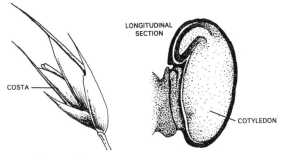

Figure 312

LONGITUDINAL SECTION

Figure 313

Crenation. A rounded projection or tooth along the margin of a leaf. Figure 318.

Crenature. See **crenation**.

Crenulate. With very small rounded teeth along the margin. Figure 319.

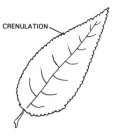

CRENULATION

Figure 319

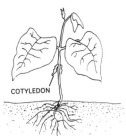

COTYLEDON

Figure 314

Figure 315

Crenulation. A very small rounded tooth along a margin; a minute crenation. Figure 319.

Creosote. An oily liquid with a strong, penetrating odor.

Crescentic. Crescent-shaped. Figure 320.

Crest. An elevated ridge or rib on a surface. Figure 321; a tuft of short, stiff hairs. Figure 322.

Crested. With a crest, usually on the back or at the summit. Figures 321 and 322.

Cribriform. Sieve-like. Figure 323.

Crampon. An adventitious root serving as a support, as in ivy. Figure 316.

Crateriform. Bowl-shaped. Figure 315.

Creeping. Growing along the surface of the ground, or just beneath the surface, and producing roots, usually at the nodes. Figure 317.

Cremocarp. See **schizocarp**.

Crena. See **crenation**.

Crenate. With rounded teeth along the margin. Figure 318.

CRAMPON

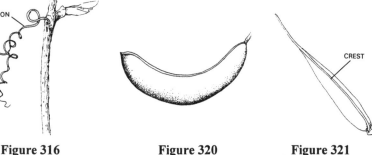

Figure 316

Figure 320

CREST

Figure 321

CREST

Figure 322

Figure 323

Cribrose. See **cribriform**.

Cribrous. See **cribriform**.

Crinite. With tufts of long, soft hairs. Figure 324.

Crinkled. Flattened and somewhat twisted, kinked, or curled. Figure 325.

Crispate. See **crisped**.

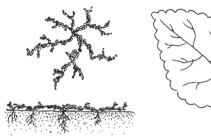

CRENATION

Figure 317

Figure 318

Crisped. Curled, wavy or crinkled. Figure 326.
Cristate. With a terminal tuft or crest. Figure 327.

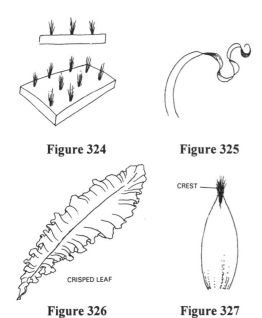

Figure 324   **Figure 325**

Figure 326   **Figure 327**

Cristulate. With a small terminal tuft or crest.
Crosier. The curled top of a young fern frond. Figure 328.
Crown. The persistent base of a herbaceous perennial. Figure 329; the top part of a tree. Figure 330; a corona. Figure 331.

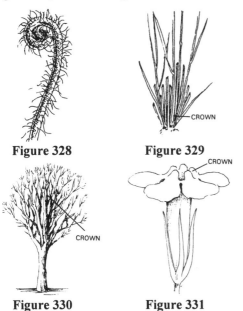

Figure 328   **Figure 329**

Figure 330   **Figure 331**

Cruciate. See **cruciform**.
Cruciform. Cross-shaped. Figure 332.
Crustaceous. Dry and brittle.
Crustose. Hard and brittle.
Cryptanthous. With the flower hidden; cleistogamous.

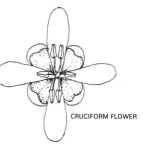

CRUCIFORM FLOWER

Figure 332

Cryptogam. A plant that does not produce seeds. (compare **phanerogam**)
Cryptophyte. See **cryptogam**.
Ctenoid. See **pectinate**.
Cucullate. Hooded or hood-shaped. Figure 333.
Cucullus. A hood. Figure 333; a seed covering external to the seed coat. Figure 334.
Cucumiform. Cucumber-shaped; cylindrical with rounded ends. Figure 335.
Culm. A hollow or pithy stalk or stem, as in the grasses, sedges, and rushes. Figure 336.

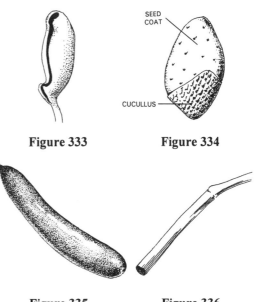

Figure 333   **Figure 334**

Figure 335   **Figure 336**

Cultivar. A form of plant originating under cultivation.
Cultrate. Shaped like a knife blade. Figure 337.
Cuneate. Wedge-shaped, triangular and tapering to a point at the base. Figure 338.

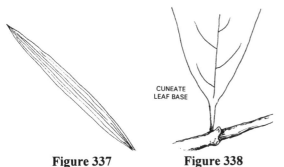

Figure 337   **Figure 338**

Cuneifoliate. With cuneate leaves.

Cuneiform. See **cuneate**.

Cupulate. Cup-shaped; with a small cup-like structure. Figure 339.

Cupule. A cup-shaped involucre, as in an acorn. Figures 340 and 341.

Cupuliform. Cup-shaped. Figure 339.

Cusp. A short, sharp, abrupt point, usually at the tip of a leaf or other organ. Figure 342.

Cuspidate. Tipped with a short, sharp, abrupt point (cusp). Figure 342.

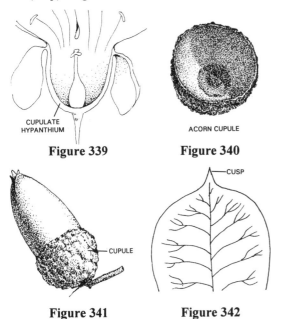

Figure 339   **Figure 340**

Figure 341   **Figure 342**

Cuticle. The waxy layer on the surface of a leaf or stem.

Cyathiform. With the form of a cyathium; cup-shaped.

Cyathium (pl. **cyathia**). The inflorescence in the genus *Euphorbia*, consisting of a cup-like invol-

ucre containing a single pistil and male flowers with a single stamen. Figure 343.

Cyclic. Occurring in apparent cycles or whorls. Figure 344.

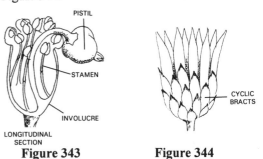

Figure 343   **Figure 344**

Cylindric. Cylinder-shaped; elongate and round in cross section. Figure 345.

Cylindrical. See **cylindric**.

Cylindroid. Shaped like a cylinder, but elliptic in cross-section. Figure 346.

Cymbiform. Boat-shaped. Figure 347.

Cyme. A flat-topped or round-topped determinate inflorescence, paniculate, in which the terminal flower blooms first. Figure 348.

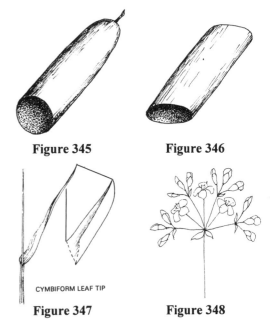

Figure 345   **Figure 346**

Figure 347   **Figure 348**

Cymose. With flowers in a cyme. Figure 348; cyme-like.

Cymule. A small cyme or a small section of a compound cyme. Figure 349.

Cynarrhodium. A fleshy, hollow fruit enclosing achenes, as the rose hip. Figure 350.

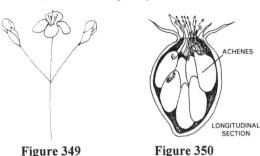

Figure 349   **Figure 350**

Cypsela. A dry, single-seeded, indehiscent fruit with an adnate calyx, as in some achenes in the Compositae (Asteraceae). Figure 351.

Dasyphyllous. With hairy or woolly leaves. Figure 352.

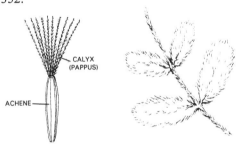

Figure 351   **Figure 352**

Deca- (prefix). Meaning ten.

Decamerous. With parts arranged in sets or multiples of ten.

Decandrous. With ten stamens.

Decantherous. With ten anthers.

Decapetalous. With ten petals.

Decaphyllous. Ten-leaved.

Decasepalous. With ten sepals.

Decaspermal. See **decaspermous**.

Decaspermous. With ten seeds.

Deciduous. Falling off, as leaves from a tree; not evergreen; not persistent.

Declinate. See **declined**.

Declined. Curved downward. Figure 353.

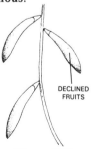

Figure 353

Decompound. More than once-compound, the leaflets again divided. Figure 354.

Decumbent. Reclining on the ground but with the tip ascending. Figure 355.

Decurrent. Extending downward from the point of insertion, as a leaf base that extends down along the stem. Figure 356.

Decurved. See **declined**.

Decussate. Arranged along the stem in pairs, with each pair at right angles to the pair above or below. Figure 357.

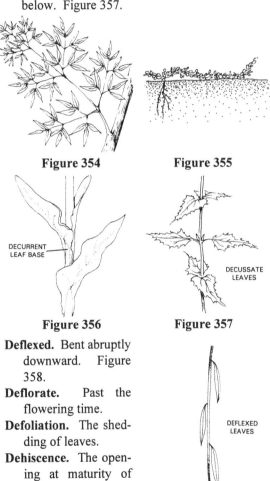

Figure 354   **Figure 355**

Figure 356   **Figure 357**

Deflexed. Bent abruptly downward. Figure 358.

Deflorate. Past the flowering time.

Defoliation. The shedding of leaves.

Dehiscence. The opening at maturity of fruits and anthers.

Figure 358

Dehiscent. Opening at maturity or when ripe to release the contents, as a fruit or an anther. Figure 359.

Deliquescent. An irregular pattern of branching without a well defined central axis from bottom to top. Figure 360.

Deltate. See **deltoid**.

Deltoid. With the shape of the Greek letter delta; shaped like an equilateral triangle. Figures 361 and 362.

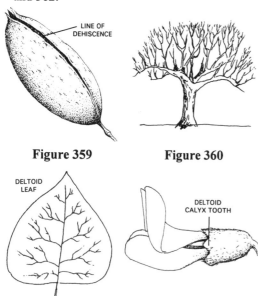

Figure 359   **Figure 360**

Figure 361   **Figure 362**

Dendriform. With a tree-like form.

Dendritic. With a branching pattern similar to that in a tree, as in some hairs in the Cruciferae (Brassicaceae). Figure 363.

Dendroid. See **dendriform**.

Dentate. Toothed along the margin, the teeth directed outward rather than forward. Figure 364.

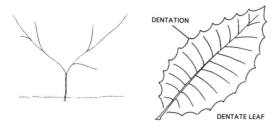

Figure 363   **Figure 364**

Dentation. The state of being dentate; a tooth along a margin, as in a leaf. Figure 364.

Denticle. A small tooth or toothlike projection. Figure 365.

Denticulate. Dentate with very small teeth. Figure 365.

Denticulation. The state of being denticulate; a denticle. Figure 365.

Dentiform. Tooth-shaped. Figure 364.

Dentoid. See **dentiform**.

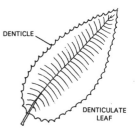

Figure 365

Denudate. Stripped bare; denuded.

Depauperate. Stunted or poorly developed, usually due to adverse environmental conditions.

Deplanate. Flattened. Figure 366.

Depressed. Flattened down from above. Figure 366.

Dermal. Of or pertaining to the epidermis.

Descending. Directed downward at a moderate angle. Figure 367.

Figure 366   **Figure 367**

Determinate. Describes an inflorescence in which the terminal flower blooms first, halting further elongation of the main axis. Figure 368.

Dextrorse. Turned to the right or spirally arranged to the right, as in the leaves on some stems. Figure 369. (compare **sinistrorse**)

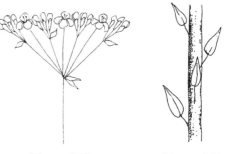

Figure 368   **Figure 369**

Di- (prefix). Meaning two or twice.

Diadelphous. Stamens united into two, often unequal, sets by their filaments. Figures 370 and 371.

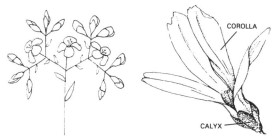

Figure 370   **Figure 371**

Dialycarpel. An ovary or fruit with separate carpels. Figure 372.

Dialycarpic. See **dialycarpous**.

Dialycarpous. With separate carpels. Figure 372.

Dialypetalous. With separate petals. Figure 373. (same as **polypetalous** and **apopetalous**)

Dialyphyllous. With the leaves distinct.

Dialysepalous. With separate sepals. Figure 374. (same as **polysepalous**)

Dialystaminous. With separate stamens. Figure 375.

Diandrous. With two stamens.

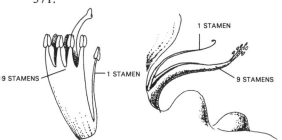

Figure 372   **Figure 373**

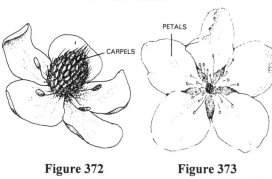

Figure 374   **Figure 375**

Diaphanous. Translucent.

Dicarpellary. See **bicarpellate**.

Dicarpellate. See **bicarpellate**.

Dichasium. A cymose inflorescence in which each axis produces two opposite or subopposite lateral axes. Figure 376.

Dichlamydeous. With two types of perianth whorls, i.e., calyx and corolla. Figure 377.

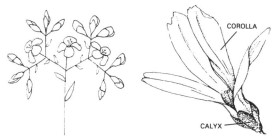

Figure 376   **Figure 377**

Dichogamic. See **dichogamous**.

Dichogamous. With the pistils and stamens maturing at different times to prevent self-fertilization. (compare **homogamous**)

Dichogamy. Having dichogamous flowers.

Dichotomous. Branched or forked into two more or less equal divisions. Figure 378.

Diclinous. With the stamens and pistils in separate flowers; imperfect. Figure 379.

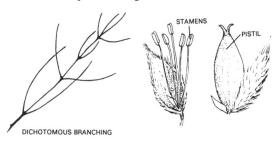

DICHOTOMOUS BRANCHING

Figure 378   **Figure 379**

Dicotyledonous. With two cotyledons. Figure 380.

Dicyclic. With two whorls.

Didymous. Developing or occurring in pairs; twin. Figure 381.

Didynamous. With two pairs of stamens of

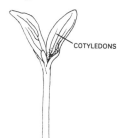

Figure 380

unequal length; occurring in pairs. Figure 382.

Diecious. See **dioecious**.

Diffuse. Widely or loosely spreading. Figure 383.

Digitate. Lobed, veined, or divided from a common point, like the fingers of a hand. Figure 384. (same as **palmate**)

Digitation. A digit-like lobe or division. Figure 384.

DIDYMOUS FRUIT

Figure 381

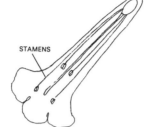

STAMENS

Figure 382

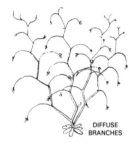

DIFFUSE BRANCHES

Figure 383

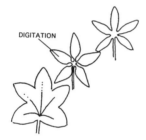

DIGITATION

Figure 384

Digitiform. Finger-like.

Digonous. With two angles, as the stems of some cacti. Figure 385.

Digynous. With two pistils. Figure 386.

Figure 385

PISTILS

Figure 386

Dilated. Flattened or expanded. Figure 387.

Dimerous. With parts arranged in sets or multiples of two.

Dimidiate. Divided unequally into halves, so that one half is so reduced as to appear lacking.

Figure 388.

Dimorphic. With two forms.

Dimorphous. See **dimorphic**.

Dioecious. Flowers imperfect, the staminate and pistillate flowers borne on different plants. (compare **monoecious**)

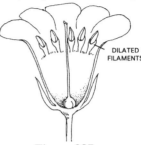

DILATED FILAMENTS

DIMIDIATE INVOLUCEL

Figure 387   **Figure 388**

Dioicous. See **dioecious**.

Dipetalous. See **bipetalous**.

Diploid. With two full sets of chromosomes in each cell.

Diplostemonous. With two series of stamens, the outer series opposite the sepals and the inner series opposite the petals; with twice as many stamens as petals. Figure 389.

Dipterous. With two wings. Figure 390.

DIPLOSTEMONOUS FLOWER

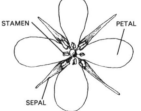

STAMEN   PETAL

SEPAL

WINGS

Figure 389   **Figure 390**

Disarticulating. Separating at maturity at a joint. Figure 391.

Disc. See **disk**.

Disciform. In the form of a disk. Figure 392.

Discoid. Resembling a disk. Figures 392 and 393; with disk flowers, as in an involucrate head of

DISARTICULATING FLOWERS

Figure 391

the Compositae (Asteraceae) which lacks ray flowers. Figure 394.

Disjunct. Occurring in widely separated geographic areas.

Disk. An enlargement or outgrowth of the receptacle around the base of the ovary; in the Compositae (Asteraceae) the central portion of the involucrate head bearing tubular or disk flowers. Figure 395.

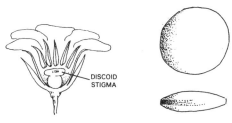

Figure 392   **Figure 393**

Figure 394   **Figure 395**

Disk flower. A regular flower of the Compositae (Asteraceae). Figure 396. (compare **ray flower**)

Dispermous. Two-seeded.

Dissected. Deeply divided into many narrow segments. Figure 397.

Figure 396   **Figure 397**

Dissepiment. See **septum**.

Dissilient. Bursting open or apart at maturity, as in some ripe fruits. Figure 398.

Distal. Toward the tip, or the end of the organ opposite the end of attachment. Figure 399. (compare **proximal**)

Distended. Stretched or swollen.

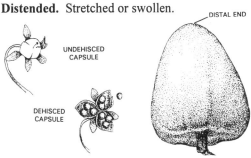

Figure 398   **Figure 399**

Distichous. In two vertical ranks or rows on opposite sides of an axis; two-ranked. Figure 400.

Distinct. Separate; not attached to like parts. Figure 401. (compare **connate**)

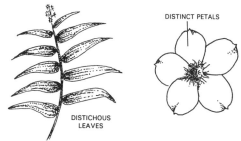

Figure 400   **Figure 401**

Dithecal anthers. Anthers lacking septi between the loculi, so that there are only two anther cells. Figure 402.

Diurnal. Occurring or opening in the daytime.

Divaricate. Widely diverging or spreading apart. Figure 403.

Divergent. Diverging or spreading. Figure 404.

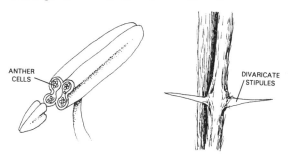

Figure 402   **Figure 403**

Divided. Cut or lobed to the base or to the midrib. Figure 405.

Dodeca- (prefix). Meaning twelve.

Dodecagynous. With twelve pistils or styles.

Dodecamerous. With parts arranged in sets or multiples of twelve.

Dodecandrous. With twelve stamens.

Dolabriform. Ax-shaped or cleaver-shaped; pick-shaped; attached at some point other than the base, usually near the middle. Figure 406.

Dorsal. Pertaining to the back or outward surface of an organ in relation to the axis, as in the lower surface of a leaf; abaxial. Figure 407. (compare **ventral**)

Dorsiventral. Having an upper and a lower surface; flattened from top to bottom. Figure 410.

Dorsoventral. See **dorsiventral**.

Double. Having a larger number of petals than usual.

Double-serrate. See **biserrate**.

Figure 410

Downy. Covered with soft, fine hairs. Figure 411.

Drepaniform. See **falcate**.

Drooping. Bending or hanging down. Figure 412.

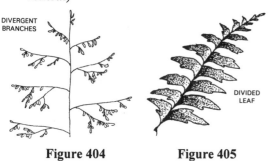

Figure 404   **Figure 405**

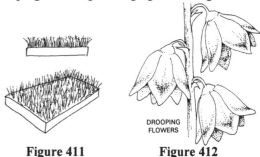

Figure 411   **Figure 412**

Drupaceous. Bearing drupes; resembling a drupe or consisting of drupes.

Drupe. A fleshy, indehiscent fruit with a stony endocarp surrounding a usually single seed, as in a peach or cherry. Figure 413.

Drupecetum. An aggregate fruit composed of many coalesced drupelets. Figure 414.

Drupel. See **drupelet**.

Drupelet. A small drupe, as in the individual segments of a raspberry fruit. Figure 414.

Drupeole. A little drupe.

Drupiferous. Bearing drupes.

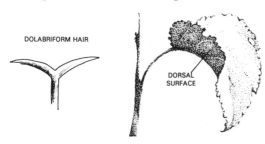

Figure 406   **Figure 407**

Dorsiferous. Borne on the back. Figure 408.

Dorsifixed. Attached at the back. Figure 409. (compare **versatile** and **basifixed**)

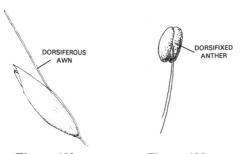

Figure 408   **Figure 409**

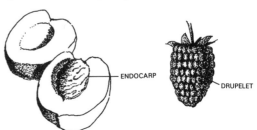

Figure 413   **Figure 414**

Dumetose. Bush-like in form.

Dumose. Full of bushes; bush-like.

Dyad. A group of two. Figure 415.

E- (prefix). Meaning without, or from, or away from.

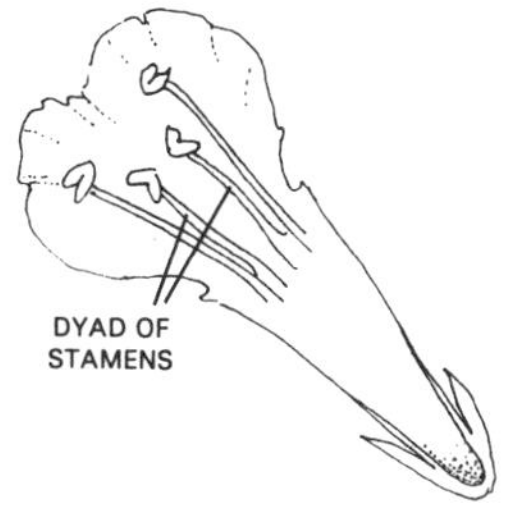

Eared. See **auriculate**.

Ebeneous. Ebony in color; black.

Ebracteate. Without bracts.

Ebracteolate. Without bracteoles.

Eburneous. Ivory-white.

Eccentric. Off-center; not positioned directly on the central axis. Figure 416.

Echinate. With prickles or spines. Figure 417.

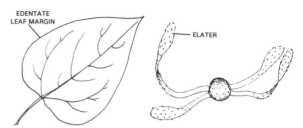

DYAD OF STAMENS

Figure 415

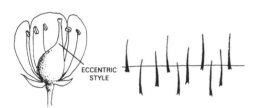

ECCENTRIC STYLE

Figure 416   **Figure 417**

Echinulate. With very small prickles or spines. Figure 418.

Ecostate. Without a midrib.

Ecotype. Those individuals adapted to a specific environment.

Ectocarp. See **exocarp**.

Edaphic. Due to, or pertaining to, the soil.

Figure 418

Edentate. Without teeth. Figure 419.

Efflorescence. The production of flowers; the period of flowering.

Effuse. See **patulous**.

Eglandular. Without glands.

Elaminate. Without a blade.

Elater. Structures attached to spores to aid in dispersal. Figure 420.

Ellipsoid. A solid body elliptic in long section and circular in cross section. Figure 421.

Elliptic. In the shape of an ellipse, or a narrow oval; broadest at the middle and narrower at the two equal ends. Figure 422.

EDENTATE LEAF MARGIN

ELATER

Figure 419   **Figure 420**

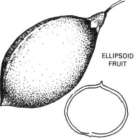

ELLIPSOID FRUIT

ELLIPTIC LEAF

Figure 421    **Figure 422**

Elliptical. See **elliptic**.

Elongate. Drawn out; lengthened.

Emarginate. With a notch at the apex. Figure 423.

Embryo. The young plant within a seed. Figure 424.

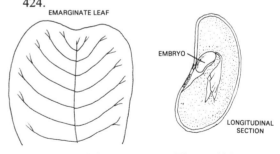

EMARGINATE LEAF

EMBRYO

LONGITUDINAL SECTION

Figure 423   **Figure 424**

Embryo sac. The megagametophyte within the ovule of a flowering plant. Figure 425.

Emergent. See **emersed**.

Emersed. Rising from, or standing out of, water.

Enation. A projection or outgrowth from the surface of an organ or structure. Figure 426.

Endemic. Peculiar to a specific geographic area or edaphic type.

Endocarp. The inner layer of the pericarp of a fruit. Figure 427. (compare **mesocarp** and **exocarp**)

Endogenous. Growing from, or originating from, within.

Endosperm. The nutritive tissue surrounding the embryo of a seed derived from the fusion of a sperm cell with the polar nuclei of the embryo sac. Figure 428.

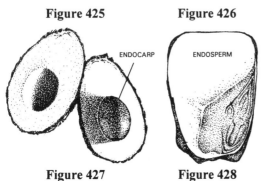

LONGITUDINAL SECTION

EMBRYO SAC

ENATION

Figure 425   **Figure 426**

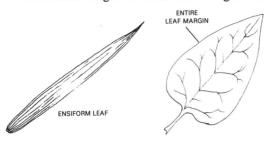

ENDOCARP   ENDOSPERM

Figure 427   **Figure 428**

Ensiform. Sword-shaped, as an *Iris* leaf. Figure 429.

Entire. Not toothed, notched, or divided, as the continuous margins of some leaves. Figure 430.

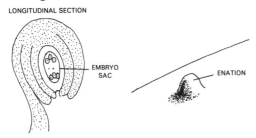

ENTIRE LEAF MARGIN

ENSIFORM LEAF

Figure 429   **Figure 430**

Entomophagous. Insect-eating; insectivorous.

Entomophilous. Insect pollinated.

Epappose. Without a pappus. Figure 431.

Epetiolate. Without a petiole, as in a sessile leaf. Figure 432.

Epetiolulate. Without a petiolule, as in a sessile leaflet. Figure 433.

Ephemeral. Lasting a very short time.

Epi- (prefix). Meaning upon.

Epiblast. A small flap of tissue on the embryo of some members of the grass family.

Epicalyx. An involucre which resembles an outer calyx, as in *Malva*. Figure 434.

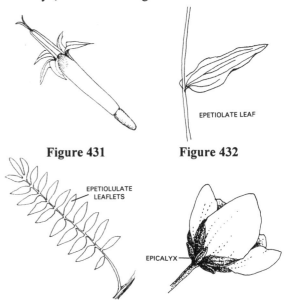

EPETIOLATE LEAF

Figure 431   **Figure 432**

EPETIOLULATE LEAFLETS

EPICALYX

Figure 433   **Figure 434**

Epicarp. See **exocarp**.

Epicotyl. That portion of the embryonic stem above the cotyledons. Figure 435.

Epidermis. The outermost cellular layer of a non-woody plant organ. Figure 436.

Epigaeal. See **epigeous**.

Epigaeous. See **epigeous**.

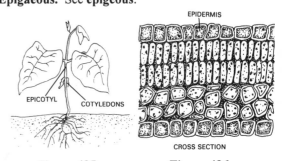

EPIDERMIS

EPICOTYL   COTYLEDONS

CROSS SECTION

Figure 435   **Figure 436**

Epigenous. Growing on the surface, as a fungus growing on the surface of a leaf.

Epigeous. Growing near the ground; said of a seedling which raises its cotyledons above the ground. Figure 437.

Epigynous. With stamens, petals, and sepals attached to the top of the ovary, the ovary inferior to the other floral parts. Figure 438.

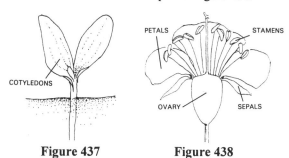

Figure 437   Figure 438

Epigyny. An epigynous condition.

Epipetalous. Attached to the petals. Figure 439.

Epipetric. Growing on a rock.

Epiphyte. A plant which grows upon another plant but does not draw food or water from it. (compare **parasite**)

Episepalous. Attached to the sepals.

Epistemonous. Attached to the stamens.

Equilateral. With sides of equal shape and length. Figure 440.

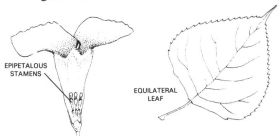

Figure 439   Figure 440

Equinoctial. With flowers that open regularly at a particular hour of the day.

Equisetoid. Resembling *Equisetum*.

Equitant. Overlapping or straddling in two ranks, as the leaves of *Iris*. Figure 441.

Eramous. With unbranched stems. Figure 442.

Erect. Vertical, not declining or spreading. Figure 443.

Erose. With the margin irregularly toothed, as if gnawed. Figure 444.

Erosulate. More or less erose.

Esculent. Edible.

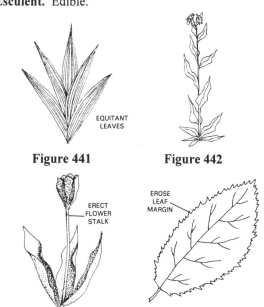

Figure 441   Figure 442

Figure 443   Figure 444

Estipellate. Without stipels. Figure 445.

Estipitate. Without a stipe. Figure 446.

Estipulate. Without stipules. Figure 445.

Estival. See **aestival**.

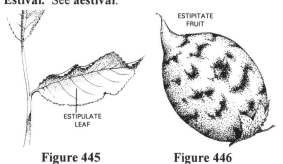

Figure 445   Figure 446

Estivation. See **aestivation**.

Etiolated. White due to a lack of chlorophyll.

Eu- (prefix). Meaning true or real.

Evanescent. Fleeting; remaining only a very short time.

Even-pinnate. Pinnately compound with a terminal pair of leaflets or a tendril rather than a single terminal leaflet, so that there is an even number of leaflets. Figures 447 and 448.

Evergreen. Having green leaves through the winter;

not deciduous.

Ex- (prefix). Same as **e-**.

Exalbuminous. Without albumen.

Exasperate. Roughened with short, stiff points. Figure 449.

Excavated. Hollowed out or concave, as the surface of some seeds. Figure 450.

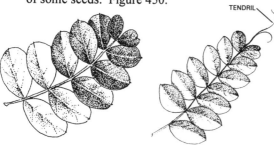

Figure 447   **Figure 448**

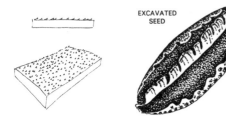

Figure 449   **Figure 450**

Excentric. See **eccentric**.

Excrescence. An irregular growth or protuberance. Figure 451.

Excurrent. Extending beyond the apex, as the midrib in some leaves. Figure 452; extending beyond what is typical, as in a leaf base which extends down the stem. Figure 453; with a prolonged main axis from which lateral branches arise. Figure 454.

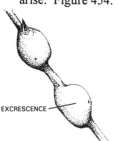

Figure 451   **Figure 452**

Excurved. Curving outward, away from the axis. Figures 455 and 456.

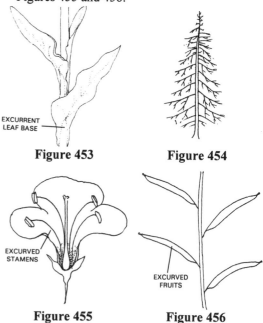

Figure 453   **Figure 454**

Figure 455   **Figure 456**

Exfoliate. To peel off in flakes or layers, as the bark of some trees. Figure 457.

Exine. The outer layer of the two-layered wall of a pollen grain. Figure 458.

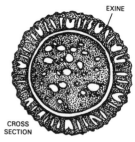

Figure 457   **Figure 458**

Exocarp. The outer layer of the pericarp of a fruit. Figure 459. (compare **mesocarp** and **endocarp**)

Exogenous. Growing from, or originating from, without.

Exomorphic. Pertaining to the external form.

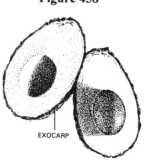

Figure 459

Exotic. Not native; introduced from elsewhere, but not completely naturalized.

Explanate. Spread out flat. Figure 460.

Exserted. Projecting beyond the surrounding parts, as stamens protruding from a corolla; not included. Figure 461.

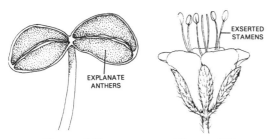

Figure 460   **Figure 461**

Exstipellate. See **estipellate**.

Exstipulate. See **estipulate**.

Extra- (prefix). Meaning outside or beyond.

Extra-axillary. Outside of but close to the axil.

Extrafloral. Outside of the flower.

Extrastaminal. Outside of the stamens.

Extrorse. Turned outward, away from the axis; opening outward. Figure 462. (compare **introrse**)

Exudate. A substance exuded or excreted from a plant.

Faboid. Bean-like.

Faceted. With many plane surfaces, like a cut gem, as in some seeds. Figure 463.

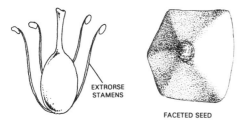

Figure 462   **Figure 463**

Facial. Pertaining to or on the face, rather than the sides or edges.

Falcate. Sickle-shaped; hooked; shaped like the beak of a falcon. Figure 464.

Falciform. See **falcate**.

Falls. The sepals of an *Iris*. Figure 465.

Farinaceous. Mealy in texture; starchy. Figure 466.

Farinose. Covered with a mealy, powdery substance. Figure 467.

Fasciated. Compressed into a bundle or band; grown closely together; with the stems malformed and flattened as if several separate stems had been fused together. Figure 468.

Fascicle. A tight bundle or cluster. Figure 469.

Fasciculate. Arranged in fascicles. Figure 469.

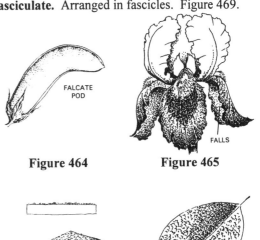

Figure 464   **Figure 465**

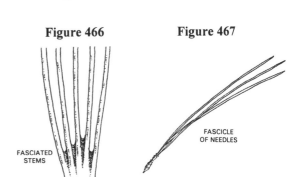

Figure 466   **Figure 467**

Figure 468   **Figure 469**

Fastigiate. Clustered, parallel, and erect, giving a broom-like appearance. Figure 470.

Faucal. Of or pertaining to the throat of a calyx or corolla.

Fauces. The throat of a calyx or corolla. Figure 471.

Faveolate. Honeycombed or pitted; alveolate. Figure 472.

Favose. See **faveolate**.

Fenestra (pl. **fenestrae**). A window-like perforation, opening, or translucent area. Figure 473.

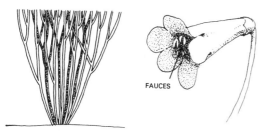

Figure 470   **Figure 471**

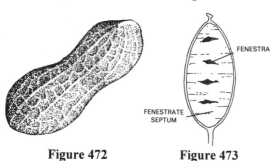

Figure 472   **Figure 473**

Fenestrate. With fenestrae. Figure 473.

Ferruginous. Rust-colored.

Fertile. Capable of bearing seeds; capable of bearing pollen.

Festucoid. Resembling the grass *Festuca*; a member of the Festucoideae group of grasses.

Fetid. With an offensive odor; stinking.

Fibriform. Fiber-like.

Fibril. A delicate fiber or hair. Figure 474.

Fibrilla. See **fibril**.

Fibrillate. See **fibrillose**.

Fibrillose. Bearing fibrils. Figure 474.

Fibrous. Bearing or resembling fibers.

Fibrous roots. A root system with all of the branches of approximately equal thickness, as in the grasses and many other monocots. Figure 475. (compare **taproot**)

Filament. A thread-like structure. Figure 476; the stalk of the stamen which supports the anther.

Figure 474

Figure 477.

Filamentose. See **filamentous**.

Filamentous. Bearing or resembling filaments. Figure 476.

Filantherous. Of a stamen with a distinct anther and filament. Figure 477.

Filical. Of or pertaining to ferns; filicoid.

Filicoid. Fern-like in appearance. Figure 478.

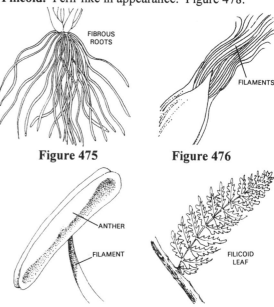

Figure 475   **Figure 476**

Figure 477   **Figure 478**

Filiferous. Bearing filaments or filament-like growths. Figure 476.

Filiform. Thread-like; filamentous. Figure 476.

Fimbria. A fringe. Figure 479.

Fimbriate. Fringed, usually with hairs or hair-like structures (fimbrillae) along the margin. Figure 479.

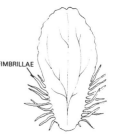

Figure 479

Fimbriation. A fringe. Figure 479.

Fimbrilla (pl. **fimbrillae**). A single unit of marginal fringe. Figure 479.

Fimbrillate. Fringed with very fine hairs. Figure 480.

Fimbriolate. See **fimbrillate**.

Fistular. See **fistulose**.

Fistulose. Hollow and cylindrical; tubular. Figure 481.

Fistulous. See **fistulose**.

Flabellate. Fan-shaped. Figure 482.

Flabelliform. See **flabellate**.

Flaccid. Limp or flabby; not rigid.

Flagellate. With long, slender runners. Figure 483.

Flagelliform. Elongate and slender; whip-like. Figure 483.

Flocculent. Bearing tufts of very fine woolly hairs; floccose. Figure 487.

Flocculose. See **flocculent**.

Floccus (pl. **flocci**). A tuft of woolly, tangled hairs. Figure 487.

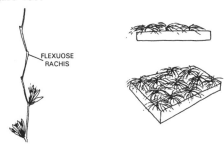

Figure 480   **Figure 481**

Figure 486   **Figure 487**

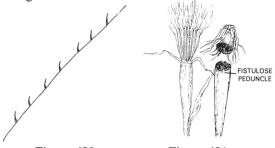

FLABELLATE STAMINODE

FLAGELLIFORM RUNNERS

Figure 482   **Figure 483**

Floral envelope. A collective term for the calyx and corolla. Figure 488; the calyx in a flower lacking a corolla. (same as **perianth**)

Floral tube. An elongated tubular portion of a perianth. Figure 489.

Florescence. The flowering period; anthesis.

Floret. A small flower; an individual flower within a dense cluster, as a grass flower in a spikelet, or a flower of the Compositae (Asteraceae) in an involucrate head. Figures 490 and 491.

Flange. A projecting rim or edge. Figure 484.

Flavescent. Yellowish.

Fleshy. Thick and pulpy; succulent. Figure 485.

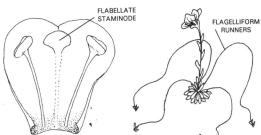

FLANGE

FLESHY LEAF

Figure 484   **Figure 485**

Flexuose. With curves or bends; sinuous; somewhat zigzagged. Figure 486.

Flexuous. See **flexuose**.

Floccose. Bearing tufts of long, soft, tangled hairs.

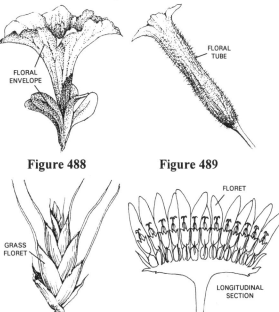

FLORAL TUBE

FLORAL ENVELOPE

Figure 488   **Figure 489**

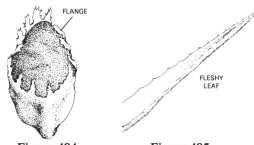

GRASS FLORET

FLORET

LONGITUDINAL SECTION

Figure 490   **Figure 491**

Flori- (prefix). Flower or flowers.

Floricane. The second-year flowering and fruiting cane (shoot) of *Rubus*. (compare **primocane**)

Florid. Flowery; covered with flowers.

Floriferous. Flower-bearing.

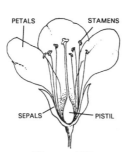

PETALS   STAMENS

SEPALS   PISTIL

Figure 492

Floscle. A flower. Figure 492.

Floscule. See **floret**.

Flower. The reproductive portion of the plant, consisting of stamens, pistils, or both, and usually including a perianth of sepals or both sepals and petals. Figure 492.

Fluted. With furrows or grooves. Figure 493.

Foliaceous. Leaf-like in color and texture; bearing leaves; of or pertaining to leaves.

Foliage. The leaves of a plant, collectively.

Foliar. Pertaining to leaves; leaf-like.

Figure 493

Foliate. Having leaves; leaf-like.

Foliated. Leaf-shaped.

Foliation. The act of producing leaves; the arrangement of leaves within a bud; foliage.

Foliature. A leaf cluster; foliage.

Foliolate. Pertaining to or having leaflets; usually used in compounds, such as **bifoliolate** or **trifoliolate**.

Foliolose. See **foliose**.

Foliose. Leafy.

Follicle. A dry, dehiscent fruit composed of a single carpel and opening along a single side, as a milkweed pod. Figure 494.

Follicular. Of or pertaining to a follicle.

Forb. A non-grasslike herbaceous plant.

Figure 494

Forcipate. Forceps-shaped. Figure 495.

Forked. Divided into two or more essentially equal branches. Figure 496.

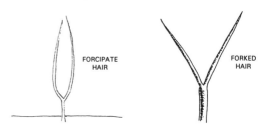

FORCIPATE HAIR

FORKED HAIR

Figure 495   Figure 496

Fornicate. Arched, as in the arched scales (fornices) in the corolla throat of many members of the Boraginaceae. Figure 497.

Fornix (pl. **fornices**). One of a set of small crests or scales in the throat of a corolla, as in many of the Boraginaceae. Figure 497.

Fovea (pl. **foveae**). A small pit or depression. Figure 498.

Foveate. With foveae; pitted. Figure 498.

Foveola (pl. **foveolae**). A little fovea; a very small pit or depression. Figure 499.

Foveolate. With foveolae; minutely pitted. Figure 499.

Free. Not attached to other organs. Figure 500.

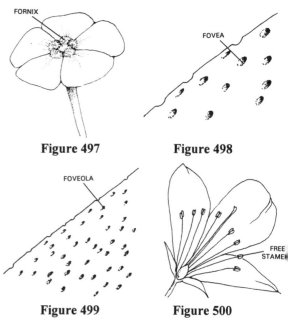

FORNIX

FOVEA

Figure 497   Figure 498

FOVEOLA

FREE STAMEN

Figure 499   Figure 500

Free-central placentation. Ovules attached to a free-standing column in the center of a unilocular ovary. Figure 501.

Fringed. With hairs or bristles along the margin. Figure 502.

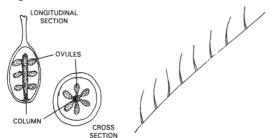

Figure 501   Figure 502

Frond. A large, divided leaf; a fern or palm leaf. Figure 503.

Frondose. With fronds; frond-like.

Fructescence. The fruiting period.

Fructiferous. Fruit-bearing.

Fructification. The fruiting process of a plant; the fruit of a plant.

Fruit. A ripened ovary and any other structures which are attached and ripen with it. Figure 504.

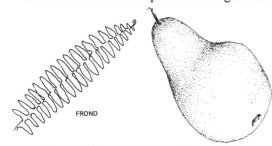

Figure 503   Figure 504

Frutescent. Shrubby or shrub-like.

Frutex. A shrub.

Frutical. See **frutescent**.

Fruticose. See **frutescent**.

Fruticulose. Somewhat shrubby; small and shrubby.

Fugacious. Falling or withering early; ephemeral. (compare **caducous**)

Fulcrum (pl. **fulcra**). A plant appendage, such as a bract, tendril, stipule, etc.

Fulvous. Tawny; dull yellowish-brown or yellowish-gray.

Funicle. See **funiculus**.

Funiculate. With a funiculus. Figure 505.

Funiculus (pl. **funiculi**). The stalk connecting the ovule to the placenta; the stalk of a seed. Figure 505.

Funnelform. Gradually widening from base to apex; funnel-shaped. Figure 506.

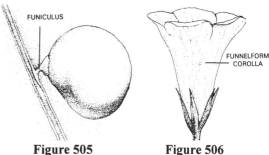

Figure 505   Figure 506

Furcate. Forked; often used in compounds, such as **bifurcate** or **trifurcate**.

Furfuraceous. Scurfy; branlike; flaky. Figure 507.

Fuscous. Dark grayish-brown; dusky.

Fusiform. Spindle-shaped; broadest near the middle and tapering toward both ends. Figure 508.

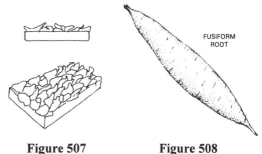

Figure 507   Figure 508

Galbulus. A cone of cypress. Figure 509.

Galea. The helmet-shaped or hood-like upper lip of some two-lipped corollas. Figure 510.

Galeate. With a galea. Figure 510; galea-like.

Galeiform. Helmet-shaped. Figure 510; galea-like.

Galericulate. See **galeate**.

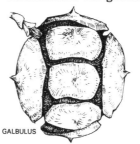
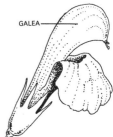

Figure 509   Figure 510

Gall. An abnormal growth caused by insects. Figure 511.

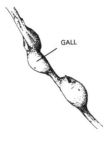

Figure 511

Gametophyte. The haploid (1n), gamete-producing generation of the plant reproductive cycle, the reduced and inconspicuous portion of the life cycle in the vascular plants. (compare **sporophyte**)

Gamo- (prefix). Meaning union of like parts.

Gamopetalous. With the petals united, at least partially. Figure 512.

Gamophyllous. With the leaves united, usually by the margins.

Gamosepalous. With the sepals united. Figure 512.

Geitonogamy. Pollination between flowers of the same plant.

Gelatinous. Jelly-like in texture.

Geminate. In equal pairs like twins. Figure 513.

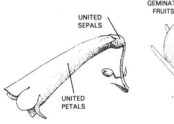

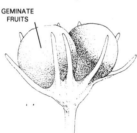

Figure 512   **Figure 513**

Gemma (pl. **gemmae**). A bud or bud-like structure, or cluster of cells which separate from the parent plant and propagate offspring plants.

Gemmate. With gemmae; reproducing by gemmae.

Gemmation. The process of reproduction by gemmae.

Gemmiferous. Producing gemmae or buds.

Gemmule. See **gemma**.

Geniculate. With abrupt knee-like bends and joints. Figure 514.

Genome. A complete chromosome set.

Gibbosity. A swelling or protuberance; the state of being gibbous. Figure 515.

Gibbous. Swollen or enlarged on one side; ventricose. Figure 515.

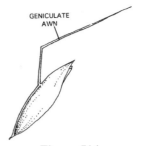

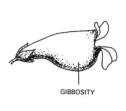

Figure 514   **Figure 515**

Glabrate. Becoming glabrous; almost glabrous.

Glabrescent. See **glabrate**.

Glabrous. Smooth; hairless.

Gladiate. Sword-shaped; ensiform. Figure 516.

Gland. An appendage, protuberance, or other structure which secretes sticky or oily substances. Figure 517.

Glandular. Of or pertaining to a gland; gland-like; bearing glands. Figure 517.

Glanduliferous. Bearing glands. Figure 517.

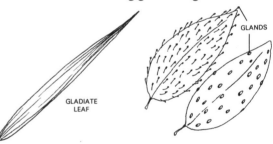

Figure 516   **Figure 517**

Glandulose. See **glandular**.

Glans. A dry, indehiscent fruit borne in a cupule, as the acorn. Figure 518.

Glaucescent. Somewhat glaucous; becoming glaucous.

Glaucous. Covered with a whitish or bluish waxy coating (bloom), as on the surface of a plum.

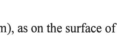

Figure 518

Globose. Globe-shaped; spherical. Figure 519.

Globular. See **globose**.

Glochid (pl. **glochidia**). A barbed hair or bristle, as the fine hairs in *Opuntia*. Figures 520 and 521.

Glochidiate. Barbed at the tip. Figure 520; bearing glochids. Figure 521.

Glome. See **glomerule**.

Glomerate. Densely clustered. Figure 522.

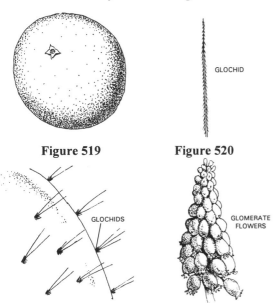

Figure 519   **Figure 520**

Figure 521   **Figure 522**

Glomerulate. Arranged in very small, dense clusters.

Glomerule. A dense cluster; a dense, head-like cyme. Figure 523.

Glumaceous. Resembling a glume; bearing glumes.

Glume. One of the paired bracts at the base of a grass spikelet. Figure 524; a chaffy bract in the grasses or sedges.

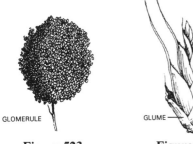

Figure 523   **Figure 524**

Glutinous. Gluey; sticky; gummy; covered with a sticky exudation.

Gorge. The throat of a flower. Figure 525.

Gossypine. Flocculent; cottony. Figure 526.

Gourd. See **pepo**.

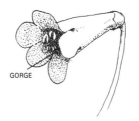 

Figure 525   **Figure 526**

Graduate. Divided or marked at regular intervals; with parts of progressively different lengths, as in some Compositae (Asteraceae) in which the outer involucral bracts are shorter than the inner. Figure 527.

Grain. A seed-like structure, as in the fruit of some *Rumex* species; a caryopsis. Figure 528.

 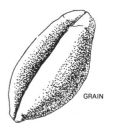

Figure 527   **Figure 528**

Granular. With small granules or grains.

Granulate. See **granular**.

Granule. A small grain.

Granuliferous. See **granular**.

Granulose. See **granular**.

Grenadine. Bright red; the color of pomegranate juice.

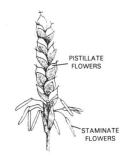

Figure 529

Gymnosperm. Plants producing seeds which are not borne in an ovary (fruit), the seeds usually borne in cones.

Gynaecandrous. With the pistillate flowers borne above the staminate, as in the inflorescence of some *Carex* species. Figure 529. (compare **androgynous**)

Gynaeceum. See **gynoecium**.

Gynaecium. See **gynoecium**.

Gynandrial. See **gynandrous**.

Gynandrium. A column bearing stamens and pistils. Figures 530 and 531.

Gynandrous. With the stamens adnate to the pistil. Figure 531.

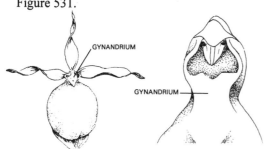

Figure 530   **Figure 531**

Gynecium. See **gynoecium**.

Gynobase. An elongation or enlargement of the receptacle, as in the flowers of the Boraginaceae. Figure 532.

Gynobasic style. A style which is attached to the gynobase as well as to the carpels. Figure 532.

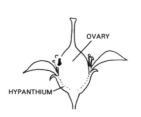

Figure 532

Gynodioecious. Having pistillate and perfect flowers on separate plants.

Gynoecium. All of the carpels or pistils of a flower, collectively. Figure 533.

Gynomonoecious. With perfect and pistillate flowers on the same plant.

Gynophore. An elongated stalk bearing the pistil in some flowers. Figure 534.

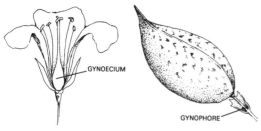

Figure 533   **Figure 534**

Gynostegium. A structure formed from the fusion of the anthers with the stigmatic region of the gynoecium, as in the Asclepiadaceae. Figure 535.

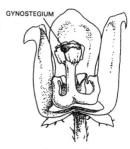

Figure 535

Gynostemial. See **gynandrous**.

Gynostemium. A structure formed from the fusion of the stamens and pistil in the Orchidaceae. Figure 531.

Habit. The general appearance, characteristic form, or mode of growth of a plant.

Habitat. The environmental circumstances or kind of place where a plant grows.

Halberd-shaped. See **hastate**.

Half-inferior. Attached below the lower half, as a flower with a hypanthium that is fused to the lower half of the ovary, giving the appearance that the other floral whorls are arising from about the middle of the ovary. Figure 536.

Halophyte. A plant which grows in salty soil.

Hamate. Hook-shaped; hooked at the tip. Figure 537.

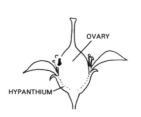

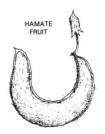

Figure 536   **Figure 537**

Hamose. See **hamate**.

Hamous. See **hamate**.

Hapaxanthic. Flowering only once.

Haplocaulis. With an unbranched stem. Figure 538.

Haplochlamydeous. See **monochlamydeous**.

Haploid. With a single

Figure 538

full set of chromosomes in each cell.

Haplopetalous. With a single series of petals. Figure 539.

Haplostemonous. With one series of stamens; with as many stamens as petals. Figure 540.

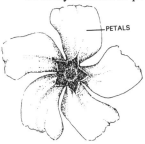

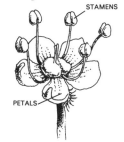

Figure 539   **Figure 540**

Hastate. Arrowhead-shaped, but with the basal lobes turned outward rather than downward; halberd-shaped. Figure 541. (compare **sagittate**)

Hastiform. See **hastate**.

Haustorium (pl. **haustoria**). A specialized root-like organ used by parasitic plants to draw nourishment from host plants.

Head. A dense cluster of sessile or subsessile flowers; the involucrate inflorescence of the Compositae (Asteraceae). Figures 542 and 543.

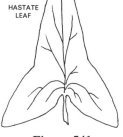

Figure 541

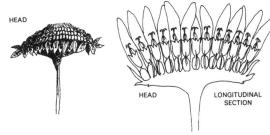

Figure 542   **Figure 543**

Heartwood. The innermost, usually somewhat darker wood of a woody stem. Figure 544.

Hebecarpous. With pubescent fruit. Figure 545.

Hebecladous. With pubescent branches. Figure 546.

Hebegynous. With pubescent pistils. Figure 547.

Hebepetalous. With pubescent petals. Figure 547.

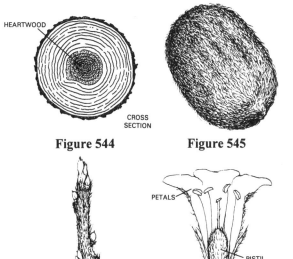

Figure 544   **Figure 545**

Figure 546   **Figure 547**

Helicoid. Coiled like a spiral or helix, as in some one-sided cymose inflorescences in the Boraginaceae. Figures 548 and 549.

Figure 548   **Figure 549**

Helmet. See **hood**.

Hemi- (prefix). Meaning half.

Hemianatropous ovule. See **hemitropous ovule**.

Hemicarp. See **mericarp**.

Hemispheric. Shaped like half of a sphere.

Hemispherical. See **hemispheric**.

Hemispheroidal. Shaped like half of a spheroid.

Hemitropous ovule. An ovule which is half inverted so that the funiculus is attached near the middle with the micropyle at a right angle. Figure 550.

Hepaticous. Liver-colored.

Heptamerous. With parts in sevens.

Herb. A plant without a persistent above-ground woody stem, the stems dying back to the ground at the end of the growing season.

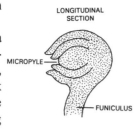

Figure 550

Herbaceous. With the characteristics of an herb; not woody.

Herbage. The non-reproductive parts of the plant; the non-woody stems, leaves, and roots of a plant.

Herbarium. A collection of dried plant specimens.

Hermaphrodite. A hermaphroditic plant.

Hermaphroditic. With pistils and stamens in the same flower; bisexual; monoclinous; perfect. Figure 551.

Hesperidium. A fleshy berrylike fruit with a tough rind, as a lemon or orange. Figure 552.

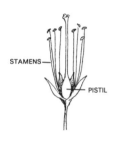

Figure 551

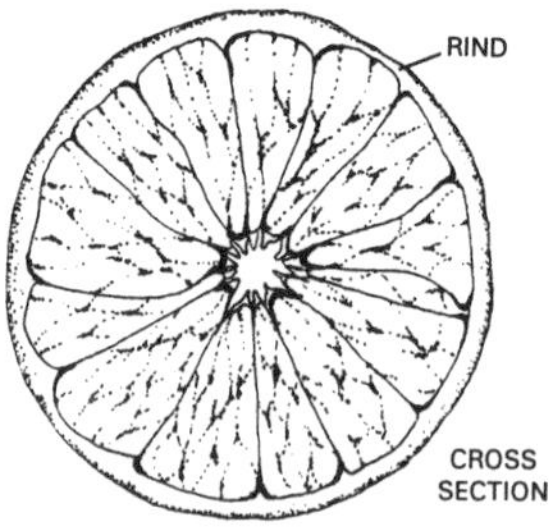

Figure 552

Heterandrous. With stamens or anthers of different forms or sizes. Figure 553.

Hetero- (prefix). Meaning different or other.

Heterocarpous. With fruit of different kinds.

Heterocephalous. With staminate and pistillate flowers in separate heads, as in some Compositae (Asteraceae).

Heterochlamydeous. With a perianth composed of distinctly different calyx and corolla whorls.

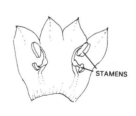

Figure 553

Figure 554.

Heterogamous. With flowers of differing sex.

Heterogamy. Having heterogamous flowers.

Heterogonous. With two or more different kinds of perfect flowers on different individuals of the same species, the kinds of flowers differing in the relative length of the pistils and stamens. (compare **homogonous**)

Heterogony. State of being heterogonous.

Heteromerous. With a variable number of parts, as in a flower with a different number of members in each floral whorl. Figure 555.

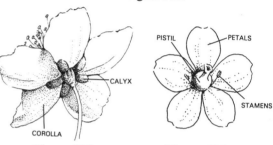

Figure 554

Figure 555

Heteromorphic. Of more than one kind or form.

Heteromorphous. See **heteromorphic**.

Heterophyllous. With different kinds of leaves on the same plant. Figure 556.

Heterosporous. Having spores of two different kinds, microspores and megaspores. Figure 557.

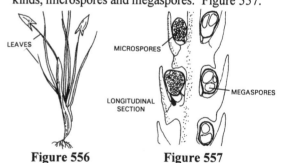

Figure 556

Figure 557

Heterostylic. With styles of different lengths in flowers of the same species. Figures 558 and 559.

Heterostylous. See **heterostylic**.

Hexa- (prefix). Meaning six.

Hexagynous. With six pistils.

Hexamerous. With parts arranged in sets or multiples of six. Figure 560.

Hexandrous. With six stamens. Figure 560.

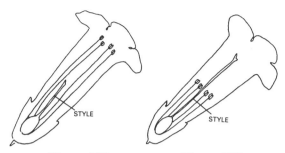

Figure 558   **Figure 559**

Hexapetalous. With six petals. Figure 560.

Hexaphyllous. With six leaves.

Hexaploid. With six full sets of chromosomes in each cell.

Hibernal. Flowering or appearing in the winter.

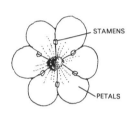

Figure 560

Hilum. A scar on a seed indicating its point of attachment. Figures 561 and 562; a scar indicating the point of attachment of the ovary in grasses.

Hip. A berry-like structure composed of an enlarged hypanthium surrounding numerous achenes, as in roses. Figures 563 and 564.

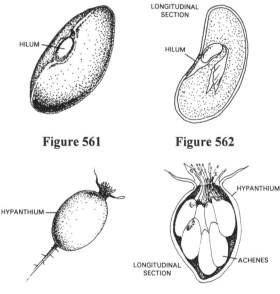

Figure 561   **Figure 562**

Figure 563   **Figure 564**

Hippocrepiform. Horseshoe-shaped. Figure 565.

Hirsute. Pubescent with coarse, stiff hairs. Figure 566.

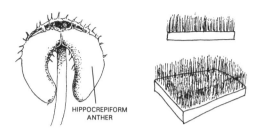

Figure 565   **Figure 566**

Hirsutulous. Pubescent with very small, coarse, stiff hairs. Figure 567.

Hirtellate. See **hirsutulous**.

Hirtellous. See **hirsutulous**.

Hispid. Rough with firm, stiff hairs. Figure 568.

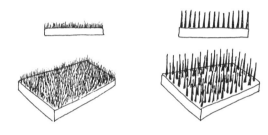

Figure 567   **Figure 568**

Hispidulous. Minutely hispid. Figure 569.

Hoary. With gray or white short, fine hairs. Figure 570.

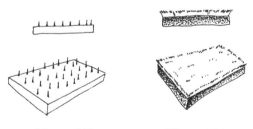

Figure 569   **Figure 570**

Holosericeous. Covered with fine, silky hairs. Figure 571.

Homo- (prefix). Meaning the same.

Homochromous. Being all of one color, as the flower heads of some Compositae (Asteraceae).

Homoeandrus. With uniform stamens.

Homogamous. With flowers of the same sex; with the pistils and stamens maturing at the same time. (compare **dichogamous**)

Homogamy. Having homogamous flowers.

HOLOSERICEOUS POD

Figure 571

Homogeneous. With parts all of the same kind; uniform.

Homogonous. With perfect flowers which do not differ in the relative length of the pistils and stamens. (compare **heterogonous**)

Homogony. State of being homogonous.

Homomorphic. All of the same kind or form.

Homomorphous. See **homomorphic**.

Homosporous. Having spores of one kind.

Homostylic. With styles of more or less constant length in flowers of the same species.

Homostylous. See **homostylic**.

Hood. A hollow, arched covering, as the upper petal in *Aconitum*. Figures 572 and 573.

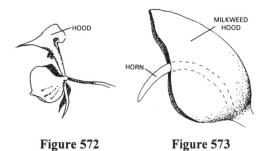

Figure 572   **Figure 573**

Hooked. Hook-shaped; bent like a hook. Figure 574.

Horn. A tapering projection resembling the horn of a cow. Figure 573.

Host. A plant providing nourishment to a parasite.

Humifuse. Spreading along the ground. Figure 575.

Humistrate. Lying on the ground. Figure 576.

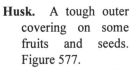

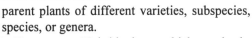

Figure 574

Figure 575

HUMISTRATE PODS

Figure 576

Husk. A tough outer covering on some fruits and seeds. Figure 577.

Hyalescent. Somewhat hyaline.

Hyaline. Thin, membranous and translucent or transparent.

Hybrid. The offspring from a cross between parent plants of different varieties, subspecies, species, or genera.

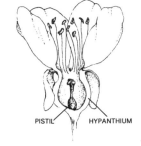

HUSK

Figure 577

Hybrid swarm. Hybrid plants which are back-crossing to the parents and crossing with themselves, so that there is a continuous intergradation of forms in the population.

Hydathode. An opening which exudes water, usually from a leaf.

Hydrophyte. A plant growing in water. (compare **mesophyte** and **xerophyte**)

Hygroscopic. Absorbing moisture from the air and sometimes swelling, shrinking, or changing position due to changes in moisture content.

PISTIL   HYPANTHIUM

Figure 578

Hypanthium. A cup-shaped extension of the floral axis usually formed from the union of the basal parts of the calyx, corolla, and androecium, commonly surrounding or enclosing the pistils. Figure 578.

Hypanthodium. An inflorescence with flowers

borne on the walls of a capitulum, as in *Ficus*. Figure 579.

Hypo- (prefix). Meaning beneath or under.

Hypochil. The basal portion of the lip of some flowers of the Orchidaceae. Figure 580.

Hypochilium. See **hypochil**.

Hypocotyl. That portion of the embryonic stem below the cotyledons. Figures 581 and 582.

Imbricate. Overlapping like tiles or shingles on a roof. Figure 585.

Immersed. Growing under water.

Imparipinnate. Odd-pinnate; unequally pinnate. Figure 586.

Imperfect. With either stamens or pistils, but not both; unisexual. Figure 587.

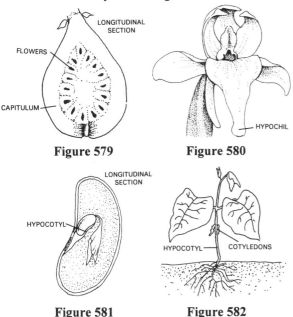

Figure 579   **Figure 580**

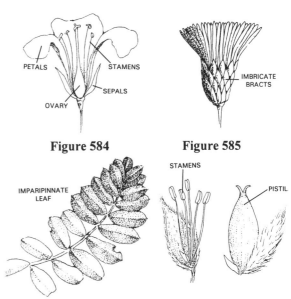

Figure 584   **Figure 585**

Figure 581   **Figure 582**

Figure 586   **Figure 587**

Hypocrateriform. Platter-shaped. Figure 583. (same as **salverform**)

Hypogaeal. See **hypogeous**.

Hypogaeous. See **hypogeous**.

Hypogeal. See **hypogeous**.

Hypogeous. Beneath the ground; said of seedling germination in which the cotyledons remain beneath the ground.

Hypogynous. With stamens, petals, and sepals attached below the ovary, the ovary superior to the other floral parts. Figure 584.

Hysteranthous. With the flowers appearing before the leaves.

Figure 583

Implicate. Twisted together; intertwined. Figure 588.

Impressed. Situated below the surface, as in some leaf veins. Figure 589; bearing marks or depressions which appear to be stamped in, as in some seeds. Figure 590.

Inaequilateral. See **inequilateral**.

Inaperturate. Lacking an aperture. Figure 591.

Incanescent. See **canescent**.

Incanous. With a whitish pubescence. Figure 592.

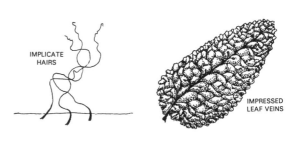

Figure 588   **Figure 589**

Incarnate. Flesh-colored.

Incised. Cut sharply, deeply, and usually irregularly. Figure 593.

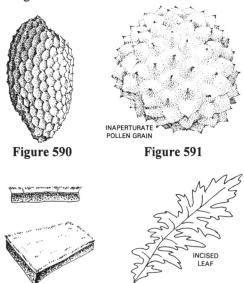

Figure 590   **Figure 591**

Figure 592   **Figure 593**

Inclined. Rising upward at a moderate angle. Figure 594.

Included. Not projecting beyond the surrounding parts, as stamens contained within a corolla; not excluded. Figure 595.

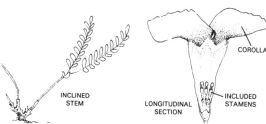

Figure 594   **Figure 595**

Incomplete. Lacking an expected part or series of parts, as in a flower lacking one of the floral whorls (i.e. sepals, petals, stamens, or pistils).

Incrassate. Thickened or swollen. Figure

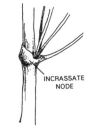

Figure 596

596.

Incumbent cotyledons. Cotyledons lying against the radicle along the back of one of the cotyledons. Figure 597. (compare **accumbent cotyledons**)

Incurved. Curved inward; curved toward the base or apex. Figure 598.

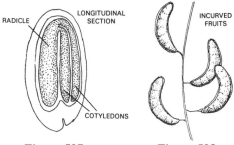

Figure 597   **Figure 598**

Indehiscent. Not opening at maturity along definite lines or by pores.

Indeterminate. Describes an inflorescence in which the lower or outer flowers bloom first, allowing indefinite elongation of the main axis. Figure 599.

Indigenous. Native to a particular area; not introduced.

Indument. The epidermal coverings of a plant, collectively.

Induplicate. With the petals or sepals edge to edge along their entire length, the margins rolled inward. Figure 600.

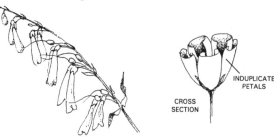

Figure 599   **Figure 600**

Indurate. Hardened.

Indusiate. With an indusium. Figure 601.

Indusium (pl. **indusia**). A thin epidermal outgrowth from a fern leaf that covers the sorus. Figure 601.

Induviate. See **marcescent**.

Inequilateral. With sides of unequal shape and

length. Figure 602.

Inermous. See **unarmed**.

Inferior. Attached beneath, as an ovary that is attached beneath the point of attachment of the other floral whorls which appear, therefore, to arise from the top of the ovary. Figure 603. (compare **superior**)

Infertile. Sterile or inviable.

Inflated. Swollen or expanded; bladdery. Figure 604.

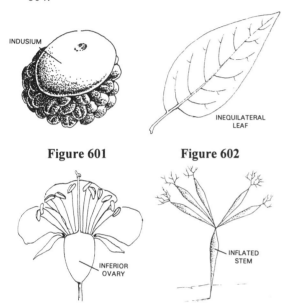

Figure 601   **Figure 602**

Figure 603   **Figure 604**

Inflexed. Bent or turned downward or inward, toward the axis. Figure 605.

Inflorescence. The flowering part of a plant; a flower cluster; the arrangement of the flowers on the flowering axis. Figure 606.

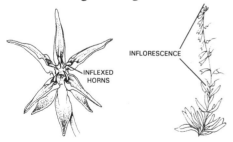

Figure 605   **Figure 606**

Infra- (prefix). Meaning below or beneath.

Infra-axillary. Below the axil.

Inframedial. Below the middle.

Infrastaminal. Below the stamens.

Infrastipular. Below the stipules.

Infundibular. See **infundibuliform**.

Infundibulate. See **infundibuliform**.

Infundibuliform. Funnel-shaped. Figure 607.

Innate. Borne at the apex, as an anther at the apex of the filament. Figure 608.

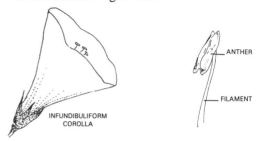

Figure 607   **Figure 608**

Innocuous. Harmless; lacking thorns or spines.

Innovation. A short, basal offset from the base of a stem. Figure 609.

Inodorous. Without an odor.

Inrolled. Curled or rolled inward. (see **involute**)

Insectivorous. Capturing and digesting insects.

Inserted. Attached to or growing out of. Figure 610.

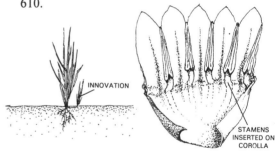

Figure 609   **Figure 610**

Insipid. Lacking taste or flavor.

Integument. The covering of the ovule which will become the seed coat. Figure 611.

Inter- (prefix). Meaning between or among.

Intercostal. Situated

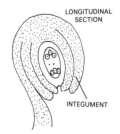

Figure 611

between the ribs or nerves. Figure 612.

Internerve. The space between two nerves. Figure 612.

Internode. The portion of a stem between two nodes. Figure 613.

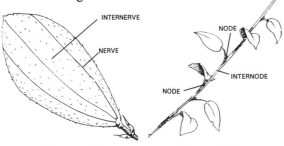

Figure 612   Figure 613

Interpetiolar. Between the petioles.

Interrupted. Not continuous.

Interruptedly pinnate. Pinnate with leaflets of various sizes intermixed. Figure 614.

Intervenous. Pertaining to the spaces between veins. Figure 612.

Intine. The inner layer of the two-layered wall of a pollen grain. Figure 615

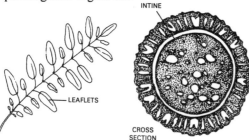

Figure 614   Figure 615

Intolerant. Not surviving well under a dense forest canopy.

Intra- (prefix). Meaning within.

Intracalycine. Within the calyx, as in the membrane within the calyx of some members of the Gentianaceae. Figure 616.

Intramarginal. Within or near the margin.

Intrastaminal. Within the androecium.

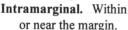

Figure 616

Intricate. Tangled together. Figure 617.

Introduced. Brought intentionally from another area; not native.

Introgression. Flow of genetic material between taxa.

Introrse. Turned inward, toward the axis; opening inward. Figure 618. (compare **extrorse**)

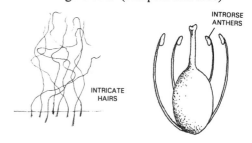

Figure 617   Figure 618

Intrusion. Protrusion into, as placentae into the cell of an ovary. Figure 619.

Intumescent. See **tumid**.

Invaginated. Sheathed; infolded. Figure 620.

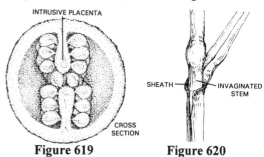

Figure 619   Figure 620

Inverted. Positioned opposite the typical direction; reversed.

Investing. Covering or surrounding. Figure 621.

Involucel. A small involucre; a secondary involucre, as in the bracts of the secondary umbels in the Umbelliferae (Apiaceae). Figure 622.

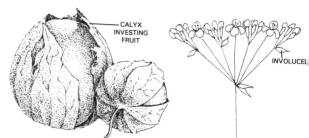

Figure 621   Figure 622

Involucellate. With an involucel. Figure 622.

Involucral. Of or pertaining to an involucre.

Involucrate. With an involucre. Figure 623.

Involucre. A whorl of bracts subtending a flower or flower cluster. Figure 623.

Involucrum (pl. **involucra**). See **involucre**.

Involute. With the margins rolled inward toward the upper side. Figure 624. (compare **revolute**)

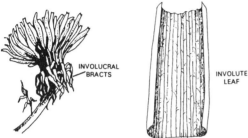

Figure 623   **Figure 624**

Iridescent. Displaying many colors, as in a rainbow.

Irregular. Bilaterally symmetrical; said of a flower in which all parts are not similar in size and arrangement on the receptacle. Figure 625. (compare **regular**, and see **zygomorphic**)

Isadelphous. With diadelphous stamens of equal number in each bundle. Figure 626.

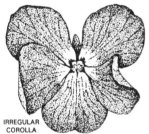

Figure 625   **Figure 626**

Isandrous. With the number of stamens equal to the number of perianth parts. Figure 627.

Isanthous. See **regular**.

Isogenous. With the same origin.

Isomerous. With an equal number of parts, as in a flower with an equal number of members in each floral

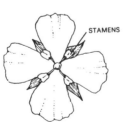

Figure 627

whorl. Figure 627.

Joint. The section of a stem from which a leaf or branch arises; a node, especially on a grass stem. Figure 628.

Jointed. Having nodes or points of articulation, as in the stems of *Opuntia*. Figure 629.

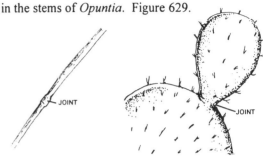

Figure 628   **Figure 629**

Jugate. With parts in pairs, as the leaflets of a pinnate leaf. Figure 630.

Julaceous. See **amentaceous**.

Junceous. See **junciform**.

Junciform. Rush-like in form. Figure 631; resembling *Juncus*.

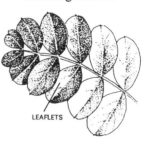
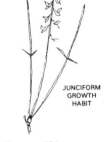

Figure 630   **Figure 631**

Karyotype. All of the chromosomes within the nucleus, especially the size, shape, and number of these chromosomes.

Keel. A prominent longitudinal ridge, like the keel of a boat. Figure 632; the two lower united petals of a papilionaceous flower. Figure 633.

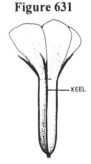

Figure 632

Keeled. Ridged, like the keel of a boat. Figure 632.

Knee. A joint or articulation, as in a grass stem. Figure 628; a bent outgrowth of a root which rises above water, as in the bald cypress. Figure 634.

Krummholz. Literally crooked forest; the low wind-contorted forest at timberline.

Label. One segment of a compound leaf. Figure 635; a labellum. Figure 636.

Labellum. Lip; the exceptional petal of an orchid blossom. Figure 636.

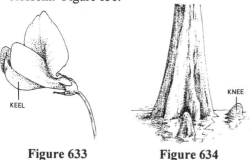

Figure 633   **Figure 634**

Figure 635   **Figure 636**

Labiate. Lipped; with parts which are arranged like lips or shaped like lips. Figures 637 and 638; of or pertaining to a member of the Labiatae (Lamiaceae).

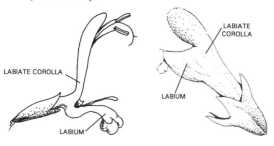

Figure 637   **Figure 638**

Labium (pl. **labia**). The lower lip of a bilabiate corolla. Figures 637 and 638.

Lacerate. Cut or cleft irregularly, as if torn. Figure 639.

Laciniate. Cut into narrow, irregular lobes or segments. Figure 640.

Lactiferous. Bearing or containing a milky latex.

Lacuna (pl. **lacunae**). An empty air space or gap within a tissue. Figure 641.

Lacustrine. Growing around lakes.

LACERATE LEAF

Figure 639

CROSS SECTION

LACINIATE LEAF   LACUNA

Figure 640   **Figure 641**

Laevigate. Lustrous; shining.

Lamella (pl. **lamellae**). An erect scale inserted on the petal in some corollas and forming part of the corona. Figure 642; a flat plate or ridge. Figure 643.

Lamellar. Of or pertaining to lamellae; with lamellae. Figure 642; plate-like. Figure 643.

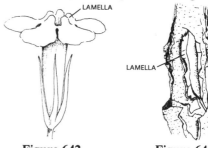

Figure 642   **Figure 643**

Lamellate. See **lamellar**.

Lamellose. See **lamellar**.

Lamina. The expanded portion, or blade, of a leaf or petal. Figures 644 and 645.

Laminar. Thin, flat, and expanded, as the blade of a leaf. Figure 644.

Laminate. With plates or layers; separating into plates or layers. Figure 643.

Lanate. Woolly; densely covered with long tangled

hairs. Figure 646.

Lanceolate. Lance-shaped; much longer than wide, with the widest point below the middle. Figure 647.

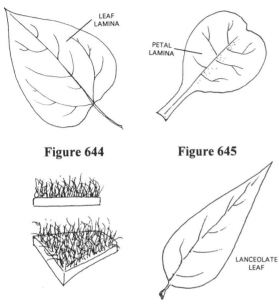

Figure 644   **Figure 645**

Figure 646   **Figure 647**

Lanose. See **lanate**.

Lanuginose. See **lanuginous**.

Lanuginous. Downy or woolly; with soft downy hairs. Figure 648.

Lanugo. A covering of soft downy hairs. Figure 648.

Lanulose. Diminutive of lanate; minutely woolly. Figure 649.

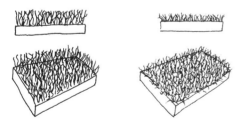

Figure 648   **Figure 649**

Lateral. Borne on or at the side. Figure 650.

Latex. A milky sap.

Laticifer. A tube or channel containing latex.

Laticiferous. Bearing or containing latex.

Latifoliate. With broad leaves. Figure 651.

Latifolious. See **latifoliate**.

Latiseptate. With a broad septum. Figure 652.

Latrorse. Dehiscing longitudinally and laterally. Figure 653.

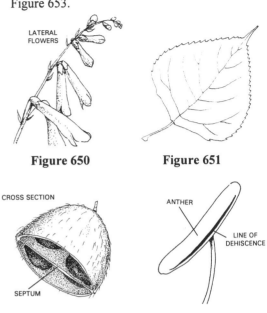

Figure 650   **Figure 651**

Figure 652   **Figure 653**

Lax. Loose; with parts open and spreading, not compact. Figure 654.

Leader. The terminal shoot of a tree or branch. Figure 655.

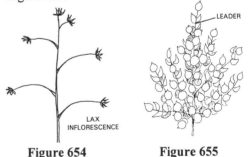

Figure 654   **Figure 655**

Leaf. An expanded, usually photosynthetic organ of a plant. Figure 656.

Leaflet. A division of a compound leaf. Figure 657.

Leaf scar. The scar remaining on a twig after a leaf falls. Figure 658.

Leaf sheath. A sheathing stipule. Figure 659.

Legume. A dry, dehiscent fruit derived from a single carpel and usually opening along two lines of dehiscence, as a pea pod. Figure 660; a plant belonging to the Leguminosae (Fabaceae) family.

Leguminous. Of or pertaining to a legume or a member of the Leguminosae (Fabaceae); legume-like.

Lemma. The lower of the two bracts (lemma and palea) which subtend a grass floret, often partially surrounding the palea. Figure 661.

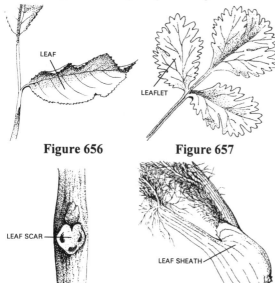

Figure 656   **Figure 657**

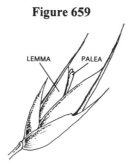

Figure 658   **Figure 659**

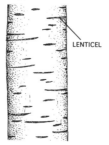

Figure 660   **Figure 661**

Figure 662

Lenticel. A slightly raised, somewhat corky, often lens-shaped area on the surface of a young stem. Figure 662.

Lenticellate. With lenticels. Figure 662.

Lenticular. Lentil-shaped (lens-

shaped); biconvex. Figure 663.

Lentiform. See **lenticular**.

Lentiginose. See **scurfy**.

Lentiginous. See **scurfy**.

Lepidoid. Scale-like.

Lepidote. Covered with small, scurfy scales. Figure 664.

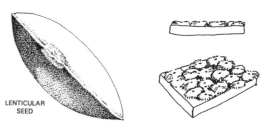

Figure 663   **Figure 664**

Liana. A woody, climbing vine.

Ligneous. Woody.

Lignified. See **ligneous**.

Lignose. See **ligneous**.

Ligula. See **ligule**.

Ligulate. With a ligule; strap-shaped. Figure 665.

Ligule. A strap-shaped organ. Figure 665; the flattened part of the ray corolla in the Compositae (Asteraceae). Figure 666; the membranous appendage arising from the inner surface of the leaf at the junction with the leaf sheath in many grasses and some sedges. Figure 667; a tongue-like projection borne at the base of the leaves above the sporangia in *Isoetes*. Figure 668.

Figure 665   **Figure 666**

Liguliform. Strap-shaped. Figure 665.

Limb. A tree branch; the expanded part of a petal or leaf. Figure 669; the expanded part of a sym-petalous corolla. Figure 670.

Limbate. Bordered, as in a leaf or flower in which one color forms an edging or margin around another. Figure 671.

Limen. A cup-like structure at the base of the androgynophore in *Passiflora*. Figure 672.

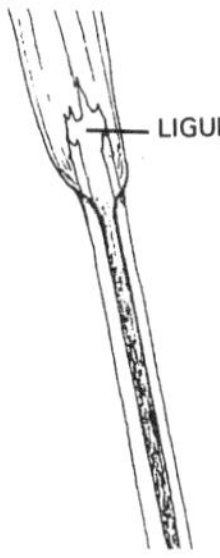

Figure 667

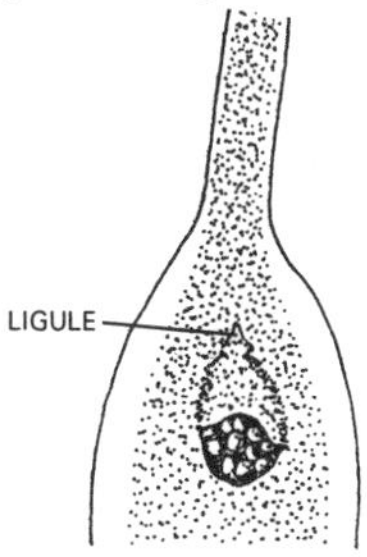

Figure 668

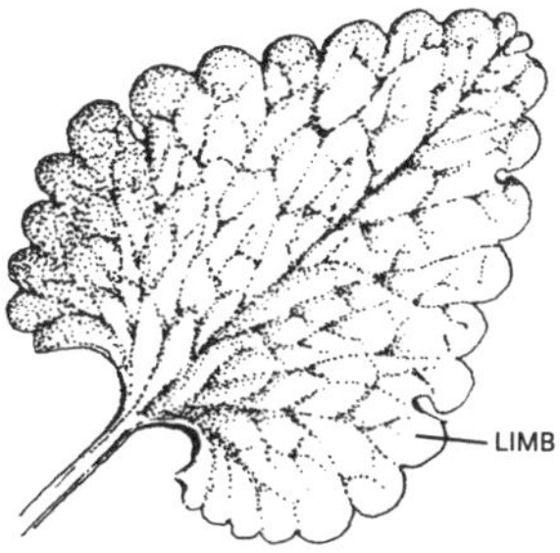

Figure 669

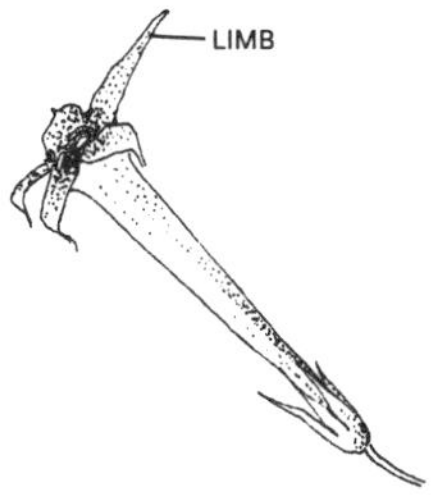

Figure 670

Figure 671

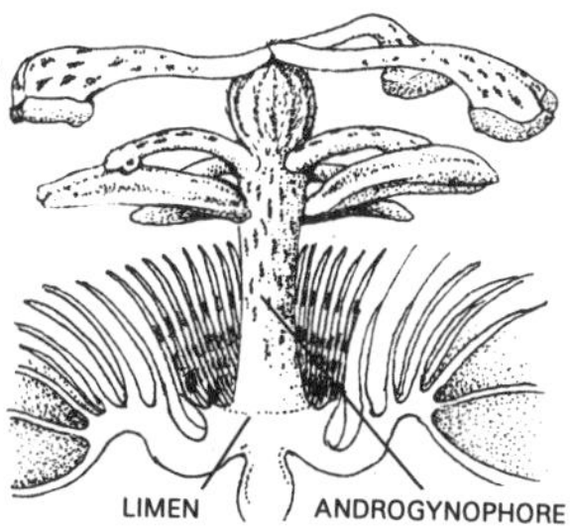

Figure 672

Linear. Resembling a line; long and narrow with more or less parallel sides. Figure 673.

Lineate. Marked with lines.

Linguiform. See **lingulate**.

Lingulate. Tongue-shaped. Figure 665.

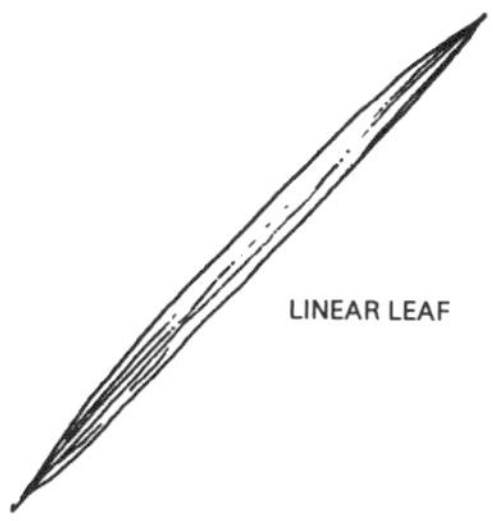

Figure 673

Lip. One of the two projections or segments of an irregular, two-lipped corolla or calyx. Figure 674; a labium; the exceptional petal of an orchid

blossom. Figure 675.

Litoral. See **littoral**.

Littoral. Growing along the shore.

Livid. Pale grayish-blue.

Lobate. In the form of a lobe; lobed. Figure 676.

Lobe. A rounded division or segment of an organ, as of a leaf. Figure 676.

Lobed. Bearing lobes which are cut less than half way to the base or midvein. Figure 676.

Lobulate. With lobules. Figure 677.

Lobule. A small lobe; a lobe-like subdivision of a lobe. Figure 677.

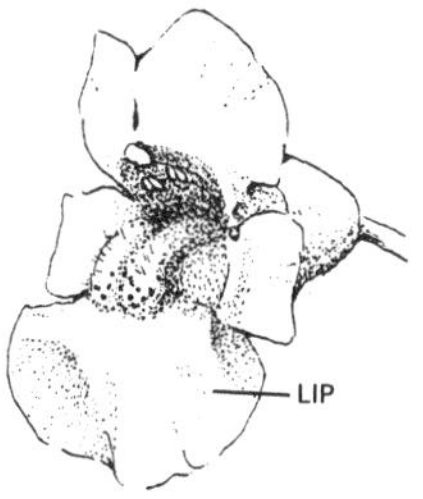

Figure 674

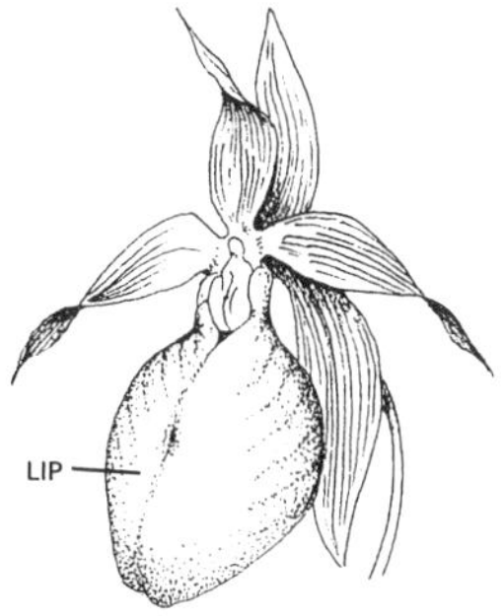

Figure 675

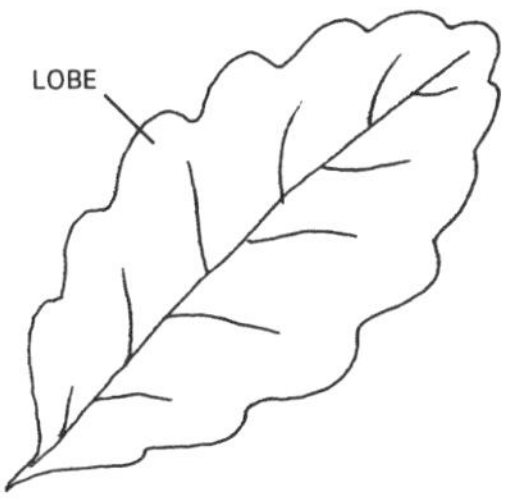

Figure 676

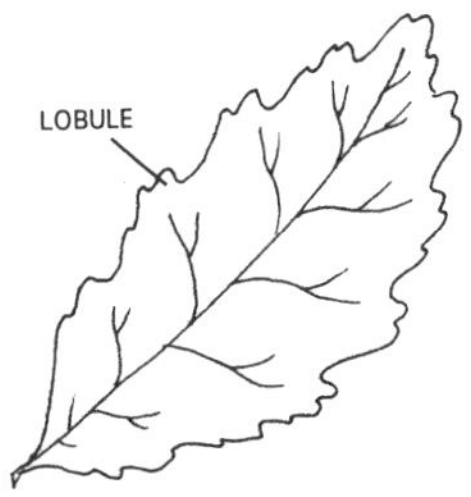

Figure 677

Locular. Of or pertaining to locules; with locules. Figure 678.

Locule. The chamber or cavity ("cell") of an organ, as in the cell of an ovary containing the seed or the pollen bearing compartment of an anther. Figure 678.

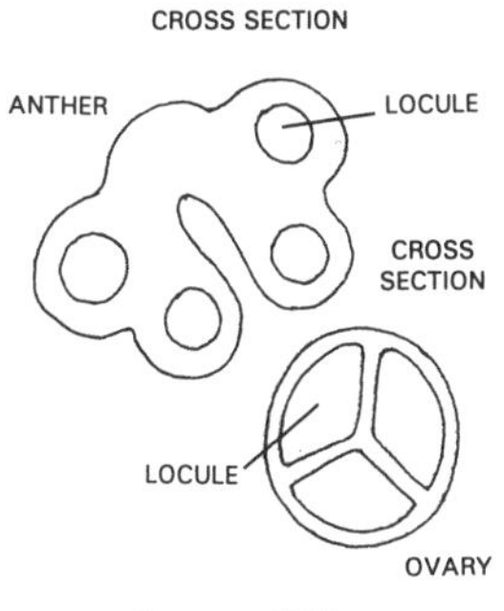

Figure 678

Loculicidal. Dehiscing through the locules of a fruit rather than through the septa. Figure 679. (compare **septicidal** and **poricidal**)

Loculus (pl. **loculi**). See **locule**.

Locusta. The spikelet of grasses. Figure 680.

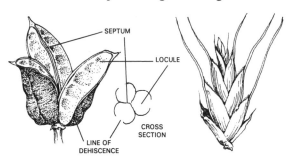

Figure 679   **Figure 680**

Lodicule. Paired, rudimentary scales at the base of the ovary in grass flowers. Figure 681.

Loment. A legume which is constricted between the seeds. Figure 682.

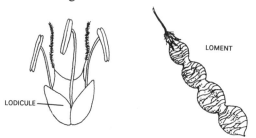

Figure 681   **Figure 682**

Lomentaceous. Loment-like; with loments.

Lomentiform. Loment-like.

Lomentum (pl. **lomenta**). See **loment**.

Longitudinal. Along the long axis of an organ. Figures 683 and 684.

Figure 683   **Figure 684**

Lorate. See **ligulate**.

Lunate. Crescent-shaped. Figure 685.

Lunulate. Diminutive of **lunate**.

Lurid. Pale brown to yellowish-brown in color.

Lustrous. Shiny or glossy.

Luteous. See **lutescent**.

Lutescent. Yellowish.

Lyrate. Lyre-shaped; pinnatifid, with the terminal lobe large and rounded and the lower lobes much smaller. Figure 686.

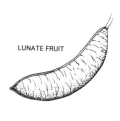

Figure 685   **Figure 686**

Lysigenous. Formed by the dissolution of tissue.

Machaerantheroid. Having involucral bracts with recurved tips. Figure 687.

Macro- (prefix). Meaning large.

Macrocladous. With long branches.

Macrophyll. The relatively large, expanded leaf of higher vascular plants. Figure 688. (compare **microphyll**)

Figure 687   **Figure 688**

Macrophyllous. With large leaves or leaflets; with macrophylls.

Macrosporangium. See **megasporangium**.

Macrospore. See **megaspore**.

Macrosporophyll. See **megasporophyll**.

Macrostylous. With a long style. Figure 689.

Macula (pl. **maculae**). A spot or blotch. Figure 690.

Figure 689

Maculate. Spotted or blotched. Figure 690.

Malacophyllous. With soft leaves.

Malodorous. Having a disagreeable odor.

Malpighiaceous hair. See **malpighian hair**.

Malpighian hair. Straight hairs tapering to both free ends and attached near the middle; pick-shaped. Figure 691. (same as **dolabriform**)

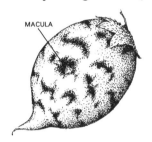

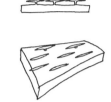

Figure 690   **Figure 691**

Malvaceous. Mallow-like.

Mammiform. Breast-shaped. Figure 692.

Mammilla (pl. **mammillae**). A nipple-like protuberance. Figure 693.

Mammillate. With nipple-like protuberances. Figure 693.

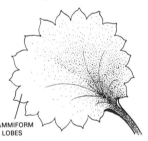

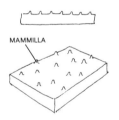

Figure 692   **Figure 693**

Mammose. See **mammillate**.

Manicate. With a thick, interwoven pubescence. Figure 694.

Many. In botanical descriptions, this term usually means more than ten.

Marcescent. Withering but persistent, as the sepals and petals in some flowers or the leaves at the base of some plants. Figure 695.

Margin. The edge, as in the edge of a leaf blade.

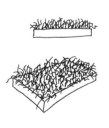

Figure 694

Figure 696.

Marginal placentation. Ovules attached to the juxtaposed margins of a simple pistil. Figure 697.

Marginate. With a distinct margin.

Maritime. Growing near the sea and often being saltwater tolerant.

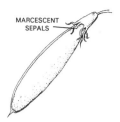

Figure 695

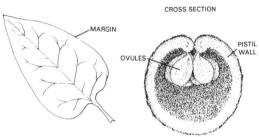

Figure 696   **Figure 697**

Masked. See **personate**.

Massula. See **pollinium**.

Mast. Nuts used for food, particularly acorns and beechnuts.

Mattula. Dense fibrous material around the petioles of palms.

Matutinal. Functioning in the morning, as in flowers which open in the morning.

Mauve. Bluish or pinkish-purple.

Mealy. With the consistency of meal; powdery, dry, and crumbly. Figure 698.

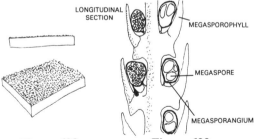

Figure 698   **Figure 699**

Medial. Of the middle; situated in the middle.

Median. See **medial**.

Mega- (prefix). Meaning large.

Megaphyllous. With large leaves.

Megasporangium (pl. **megasporangia**). A spore-producing structure (sporangium) which bears megaspores. Figure 699.

Megaspore. A female spore which will give rise to the female gametophyte. Figure 699.

Megasporophyll. A modified leaf which bears one or more megasporangia. Figure 699.

Melanophyllous. With dark leaves.

Melanoxylon. Dark wood.

Membranaceous. See **membranous**.

Membranous. Thin, soft, flexible, and more or less translucent, like a membrane.

Meniscoid. Concavo-convex; one side concave and the other convex. Figure 700.

Mentum. A projection formed by the extension of the base of the column in some orchids. Figure 701.

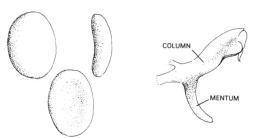

Figure 700   **Figure 701**

Mephitic. Having a strong, disagreeable odor.

Mericarp. A section of a schizocarp; one of the two halves of the fruit in the Umbelliferae (Apiaceae). Figure 702.

Meristem. Undifferentiated, actively dividing tissues at the growing tips of shoots and roots. Figure 703.

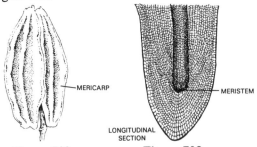

Figure 702   **Figure 703**

Meristematic. Of or pertaining to the meristem.

-merous (suffix). Meaning parts of a set. A 5-merous corolla would have five petals.

Mesic. Moist.

Meso- (prefix). Meaning middle.

Mesocarp. The middle layer of the pericarp of a fruit. Figure 704. (compare **endocarp** and **exocarp**)

Mesocotyl. That portion of the embryonic stem between the coleoptile and the scutellum in a grass embryo. Figure 705.

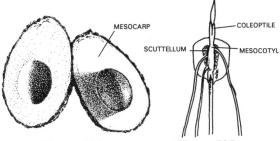

Figure 704   **Figure 705**

Mesophyll. The central tissues of a leaf between the upper and lower epidermis. Figure 706.

Mesophyte. A plant growing in average moisture conditions. (compare **hydrophyte** and **xerophyte**)

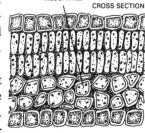

Figure 706

Metandry. Female flowers maturing before the male flowers; protogyny.

Micro- (prefix). Meaning small.

Microphyll. The relatively small, narrow, single-veined leaf of some lower vascular plants. Figure 707. (compare **macrophyll**)

Micropyle. The opening in the integuments of the ovule. Figure 708.

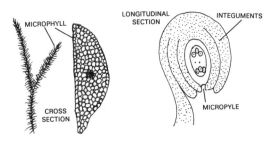

Figure 707   **Figure 708**

Microsporangium (pl. **microsporangia**). A spore-producing structure (sporangium) which bears microspores. Figure 709.

Microspore. A male spore which will give rise to the male gametophyte. Figure 709.

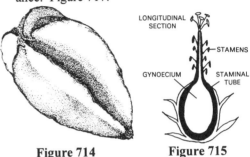

Figure 709

Microsporophyll. A modified leaf which bears one or more microsporangia. Figure 709; a stamen.

Midlobe. The central lobe. Figure 710.

Midnerve. The central nerve. Figure 711.

Midrib. The central rib or vein of a leaf or other organ. Figure 711.

Midvein. The central vein. Figure 712.

Mitriform. Shaped like a mitra. Figure 713.

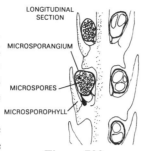

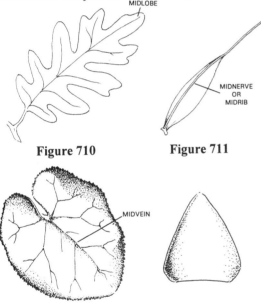

Figure 710

Figure 711

Figure 712

Figure 713

Mixed bud. A bud which produces both leaves and flowers.

Mixed inflorescence. An inflorescence with both racemose and cymose portions.

Molendinaceous. With large, wing-like developments. Figure 714.

Monad. A single individual that is free from other

such individuals rather than being united into a group.

Monadelphous. Stamens united by the filaments and forming a tube around the gynoecium. Figures 715 and 716.

Monandrous. With a single stamen.

Monanthous. One-flowered.

Monecious. See **monoecious**.

Moniliform. Cylindrical and constricted at regular intervals, giving a beaded necklace-like appearance. Figure 717.

Figure 714

Figure 715

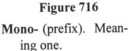

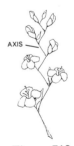

Figure 716

Figure 717

Mono- (prefix). Meaning one.

Monocarpic. Flowering and bearing fruit only once and then dying. The term may be applied to perennials, biennials, or annuals.

Monocarpous. With one carpel.

Figure 718

Monocephalous. With one head.

Monochasial. With the form of a monochasium.

Monochasium. A type of cymose inflorescence with only a single main axis. Figure 718.

Monochlamydeous. With only one type of perianth

member. Figure 719.

Monochromatic. Of a single color.

Monoclinous. With pistils and stamens in the same flower; perfect. Figure 720.

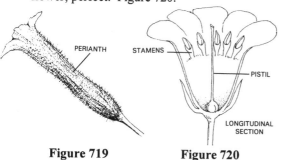

Figure 719   **Figure 720**

Monocotyledon. Plants with a single seed leaf, or cotyledon. Figure 721.

Monocotyledonous. With a single cotyledon. Figure 721.

Monocyclic. With a single whorl. Figure 722.

Monodynamous. With one stamen distinctly larger than the others. Figure 723.

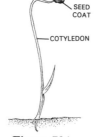

Figure 721

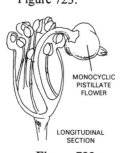

Figure 722

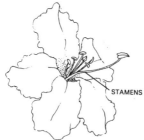

Figure 723

Monoecious. Flowers imperfect, the staminate and pistillate flowers borne on the same plant. (compare **dioecious**)

Monogynous. With one carpel. Figure 724.

Monolocular. With a single cell or cham-

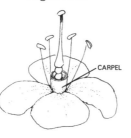

Figure 724

ber; unilocular. Figure 725.

Monomerous. With a single member, as in a floral whorl with only one part. Figure 722.

Monomorphic. With a single form; all alike in appearance.

Monopetalous. See **sympetalous** or **gamopetalous**.

Monophyllous. Of a single leaf; with simple leaves; said of plants that have simple leaves though their relatives have compound leaves.

Monopodial. Of or pertaining to a monopodium; with the branches arising from a single main axis. Figure 726.

Monopodium (pl. **monopodia**). A single main axis giving rise to lateral branches. Figure 726. (compare **sympodium**)

Monopterous. With a single wing. Figure 727.

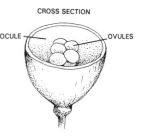

Figure 725

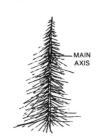

Figure 726   **Figure 727**

Monosepalous. See **gamosepalous**.

Monospermous. One-seeded.

Monostachous. With flowers arranged in a single spike.

Monostichous. In a single vertical rank or row. Figure 728.

Monostylous. With a single style. Figure 729.

Monosymmetrical. Bilaterally symmetrical; zygomorphic. Figure 730.

Monotrichous. With a single bristle. Figure 731.

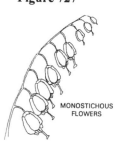

Figure 728

Figure 729

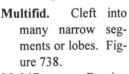

MONOSYMMETRICAL COROLLA

Figure 730

Monotypic. A taxon with only a single representative, as a genus with a single species or a family with a single genus.

Montane. A plant growing in the mountains.

Moschate. With a musky scent.

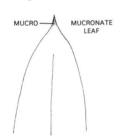

Figure 731

Mottled. With colored spots or blotches. Figure 732.

Mucilaginous. Slimy and moist; mucilage-like.

Mucro. A short, sharp, abrupt point, usually at the tip of a leaf or other organ. Figure 733.

Figure 732

Figure 733

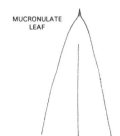

Figure 734

Figure 735

Mucronate. Tipped with a short, sharp, abrupt point (mucro). Figure 733.

Mucronulate. Tipped with a very small mucro. Figure 734.

Multi- (prefix). Meaning many.

Multiciliate. With many cilia. Figure 735.

Multicipital. Many-headed, as the crown of a root divided into a number of caudices. Figure 736.

Multicostate. With many ribs. Figure 737.

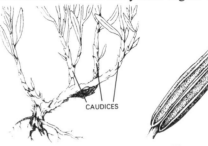

CAUDICES

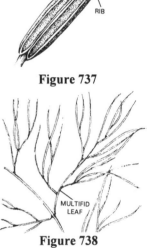

RIB

Figure 736

Figure 737

Multifid. Cleft into many narrow segments or lobes. Figure 738.

Multiflorous. Bearing many flowers.

Multifoliate. Bearing many leaves or leaflets.

Multiparous. A cyme with many lateral axes.

Multipartite. See **multifid**.

Multiperennial. See **pliestesial**.

Multiple fruit. A fruit formed from several separate flowers crowded on a single axis, as a mulberry or pineapple. Figure 739.

Multiplicate. Repeatedly folded. Figure 740.

MULTIFID LEAF

Figure 738

MULBERRIES

Figure 739

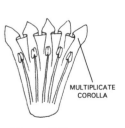

MULTIPLICATE COROLLA

Figure 740

Multiradiate. With many rays. Figure 741.

Multiseptate. With many septae or partitions. Figure 742.

Muricate. Rough with small, sharp projections or points. Figure 743.

Murication. A small, sharp projection or point. Figure 743.

Muriculate. Very finely muricate. Figure 744.

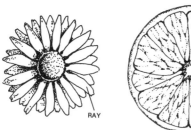

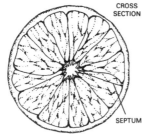

Figure 741   **Figure 742**

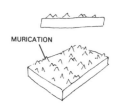

Figure 743   **Figure 744**

Muticous. Blunt, without a point or spine. Figure 745.

Mycorrhiza (pl. **mycorrhizae**). A symbiotic relationship between a fungus and the root of a plant.

Mycotrophic. Modified by a mycorrhizal relationship.

Myochrous. Mouse-colored.

Nacreous. With a pearly luster; pearlescent.

Naked. Lacking hairs, structures, or appendages typically present, as in a flower lacking a perianth; nude.

Naked bud. A bud lacking scales.

Napaceous. See **napiform**.

Figure 745

Napiform. Turnip-shaped. Figure 746.

Nascent. Beginning to develop, but not yet fully formed.

Natant. Floating in water; immersed.

Naturalized. Plants introduced from elsewhere, but now established.

Naucum. The soft, fleshy part of a drupe. Figure 747.

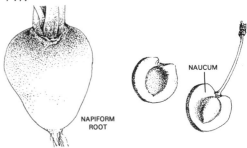

Figure 746   **Figure 747**

Nautiloid. Spiral-shaped, like a *Nautilus* shell. Figure 748.

Navicular. Boat-shaped. Figure 749.

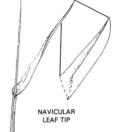

Figure 748   **Figure 749**

Nebulose. Indistinct, as in a fine, diffuse inflorescence. Figure 750.

Neck. Any constricted part of a plant, as the narrowed section of a perianth, or the point where the blade joins the sheath in some grass leaves. Figures 751 and 752.

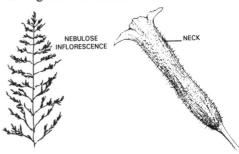

Figure 750   **Figure 751**

Nectar. A sugary, sticky fluid secreted by many plants.

Nectar gland. See **nectary**.

Nectar guides. Lines or spots, often invisible except in ultraviolet light, directing pollinators toward the nectaries. Figure 753.

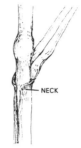

Figure 752

Nectariferous. With nectar.

Nectary. A tissue or organ which produces nectar. Figure 754.

Needle. A slender, needle-shaped leaf, as in the pinaceae. Figure 755.

Nema (pl. **nemata**). A filament or thread.

Nephroid. Kidney-shaped; reniform. Figure 756.

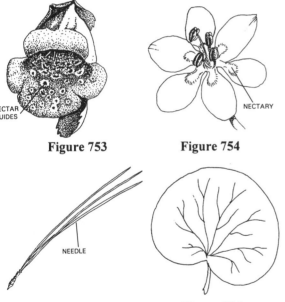

Figure 753   **Figure 754**

Figure 755   **Figure 756**

Nervation. The arrangement of nerves or veins in an organ. Figure 757.

Nerve. A prominent, simple vein or rib of a leaf or other organ. Figure 757.

Nerviform. Resembling a nerve.

Nervose. With prominent nerves. Figure 757.

Netted. See **net-veined**.

Net-veined. In the form of a network; reticulate.

Figure 757.

Neuter. Lacking functional stamens or pistils.

Neutral. See **neuter**.

Nidulent. Lying within a cavity; embedded within a pulp. Figure 758.

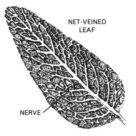

Figure 757   **Figure 758**

Nigrescent. Blackish.

Nitid. Lustrous; shining.

Niveous. White.

Nocturnal. Functioning at night, as in flowers which open at night.

Nodal. Of, on, or pertaining to a node. Figure 759.

Nodding. Bent to one side and downward. Figure 760.

Node. The position on the stem where leaves or branches originate. Figure 759.

Nodiferous. With nodes. Figure 759.

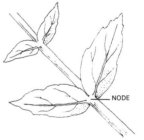

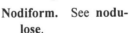

Figure 759   **Figure 760**

Nodiform. See **nodulose**.

Nodose. With knobs or nodules. Figures 761 and 762; with nodes. Figure 759.

Nodule. A swelling or knob. Figure 761.

Nodulose. With minute knobs or nodules. Figure 763.

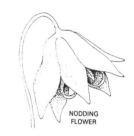

Figure 761

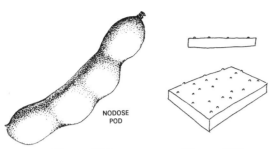

Figure 762   **Figure 763**

Nomad. A plant growing in pastures.
Nomophilous. Growing in pastures.
Notate. Marked with lines or spots. Figure 764.
Notched. With a small cut or notch. Figure 765.
Nucamentaceous. Catkin-like; indehiscent.
Nucamentum. A catkin or ament. Figure 766.
Nucellus. The part of the ovule just beneath the integuments and surrounding the female gametophyte. Figure 767.

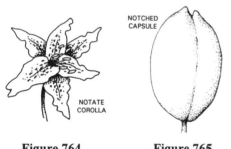

Figure 764   **Figure 765**

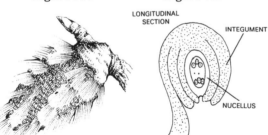

Figure 766   **Figure 767**

Nuciferous. Bearing nuts.
Nude. See **naked**.
Nudicaul. With leafless stems.
Numerous. In botanical descriptions, this term usually means more than ten.
Nut. A hard, dry, indehiscent fruit, usually with a single seed. Figure 768.
Nutant. Drooping; nodding. Figure 769.

Figure 768   **Figure 769**

Nutlet. A small nut; one of the lobes or sections of the mature ovary of some members of the Boraginaceae, Verbenaceae, and Labiatae (Lamiaceae). Figure 770.
Nux. A nut. Figure 768.
Nyctanthous. Night-flowering.

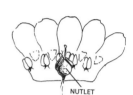

Figure 770

Nyctigamous. Opening at night.
Nyctitropic. Movement or positioning of plant organs at night that is unlike those occurring during the day.
Ob- (prefix). Meaning inversion; in a reverse direction.
Obclavate. Club-shaped, with the attachment at the broad end. Figure 771.
Obcompressed. Compressed opposite the usual way, as in a structure which is flattened dorso-ventrally when similar structures are flattened laterally. Figure 772.

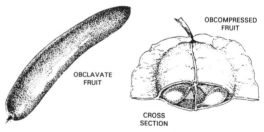

Figure 771   **Figure 772**

Obconic or **obconical.** Conical or cone-shaped, with the attachment at the narrow end. Figure 773.

Obcordate. Inversely cordate, with the attachment at the narrower end. Figure 774; sometimes refers to any leaf with a deeply notched apex. Figure 775.

Obcordiform. See **obcordate**.

Obdeltoid. Deltoid, with the attachment at the pointed end. Figure 776.

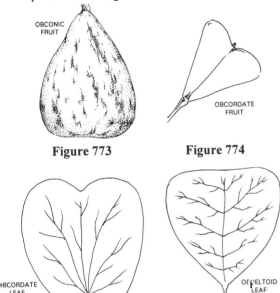

Figure 773   **Figure 774**

Figure 775   **Figure 776**

Obdiplostemonous. Having two whorls of stamens, the outer whorl opposite the petals and the inner whorl opposite the sepals. Figure 777.

Obelliptic or **obelliptical.** Almost elliptic, but with the distal end somewhat larger than the proximal end. Figure 778.

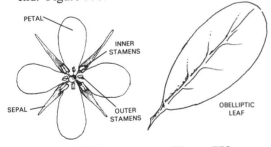

Figure 777   **Figure 778**

Oblanceolate. Inversely lanceolate, with the attachment at the narrower end. Figure 779.

Oblate. Spheroidal and flattened at the poles. Figure 780.

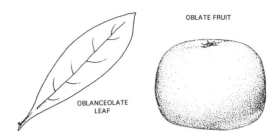

Figure 779   **Figure 780**

Obligate. Restricted to particular conditions or circumstances, as a parasite incapable of independent survival.

Oblique. With unequal sides, especially a leaf base. Figure 781; slanting.

Oblong. Two to four times longer than broad with nearly parallel sides. Figure 782.

Obovate. Inversely ovate, with the attachment at the narrower end. Figure 783.

Obovoid. Inversely ovoid, with the attachment at the narrower end. Figure 784.

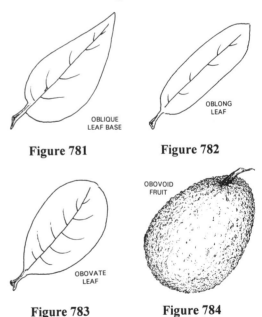

Figure 781   **Figure 782**

Figure 783   **Figure 784**

Obpyramidal. Inversely pyramidal, with the attachment at the narrower end. Figure 785.

Obsolescent. See **obsolete**.

Obsolete. An organ or structure which is much reduced and likely nonfunctional, though believed at one time to have been more perfectly formed; vestigial. Figure 786.

Obturator. A small glandular structure attached to the pollinia of members of the Asclepiadaceae and Orchidaceae which closes the opening to the anther. Figure 787.

Obtuse. Blunt or rounded at the apex; with the sides coming together at the apex at an angle greater than 90 degrees. Figure 788.

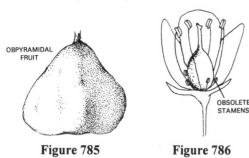

Figure 785   **Figure 786**

Figure 787   **Figure 788**

Obvolute. A vernation in which two leaves are overlapping in bud such that one half of each is external and the other half is internal. Figure 789.

Ocellus. An eye-like marking, as in a spot of color encircled by a band of another color. Figure 790.

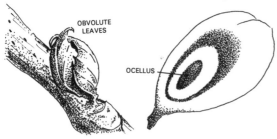

Figure 789   **Figure 790**

Ochraceous. Ochre-colored.
Ochrea (pl. **ochreae**). See **ocrea**.
Ochreate. See **ocreate**.

Ochroleucous. Yellow-ish-white; cream-colored.

Ocrea (pl. **ocreae**). A sheath around the stem formed from the stipules, as in many members of the Polygonaceae. Figure 791.

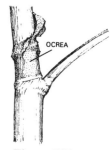

Figure 791

Ocreate. With sheathing stipules. Figure 791.

Ocreola (pl. **ocreolae**). A minute stipular sheath around the secondary divisions of the inflorescence in some members of the Polygonaceae. Figure 792.

Ocreolate. With minute sheathing stipules; often applied to bract bases. Figure 792.

Octandrous. With eight stamens. Figure 793.

Octogynous. With eight pistils or styles. Figure 793.

Octolocular. With eight locules. Figure 794.

Octopetalous. With eight petals. Figure 793.

Octoradiate. With eight ray flowers. Figure 795.

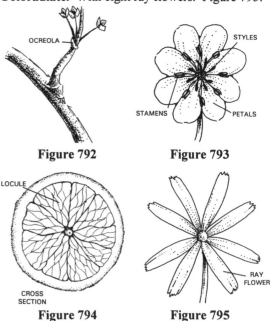

Figure 792   **Figure 793**

Figure 794   **Figure 795**

Octosepalous. With eight sepals. Figure 796.
Octostemonous. With eight stamens. Figure 793.
Octostichous. In eight ranks or rows.

Odd-pinnate. Pinnately compound with a terminal leaflet rather than a pair of leaflets or a tendril, so that there is an odd number of leaflets. Figure 797.

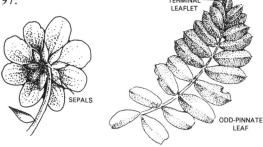

Figure 796 **Figure 797**

Odoriferous. With a distinct odor.

Offset. A short, often prostrate, shoot originating near the ground at the base of another shoot. Figure 798.

Offshoot. A shoot or branch arising from a main stem. Figure 799.

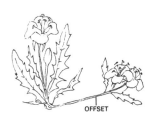 

Figure 798 **Figure 799**

Oil tube. Narrow ducts in the walls of the fruit of many members of the Umbelliferae (Apiaceae) containing volatile oils. Figure 800.

Oleaginous. Oily; oil-producing.

Oleiferous. Oil-bearing.

Oligandrous. With few stamens.

Oligocarpic. See **oligocarpous**.

Oligocarpous. Bearing less than the typical amount of fruit.

Oligomerous. With less than the typical number of parts.

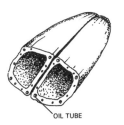

Figure 800

Oligophyllous. With few leaves.

Oligospermous. With few seeds.

Olivaceous. Olive-green; olive-like.

Operculate. With an operculum. Figure 801.

Operculum. A small lid, such as the deciduous cap of a circumscissile capsule. Figure 801.

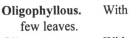

Figure 801

Opposite. Borne across from one another at the same node, as in a stem with two leaves per node. Figure 802; borne over, or on the same radius as other organs rather than between other organs, as a stamen in front of a petal. Figure 803. (compare **alternate**)

Oppositiflorous. With opposite pedicels or peduncles. Figure 804.

Oppositifolious. With opposite leaves. Figure 802.

Orbicular. Approximately circular in outline. Figure 805. (compare **spherical**)

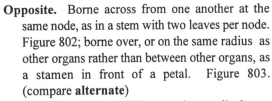

Figure 802 **Figure 803**

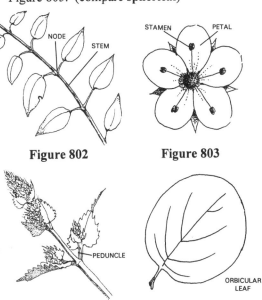

Figure 804 **Figure 805**

Orbiculate. See **orbicular**.

Orchioid. Orchid-like.

Organ. A plant part with a specific function, as a

leaf.

Orifice. An opening or mouth, as the mouth-like opening of a tubular corolla. Figure 806.

Ornithophilous. Pollinated by birds.

Orophilous. Growing in mountainous areas.

Orthocladous. With straight branches. Figure 807.

Orthopterous. Straight-winged. Figure 808.

Orthostichous. With parts arranged in straight ranks or rows. Figure 809.

Orthotropic. Of, pertaining to, or exhibiting an essentially vertical growth habit. Figure 810.

Orthotropous ovule. An ovule which is straight and erect. Figure 811.

Outcross. To transfer pollen from the anthers of the flowers of one plant to the stigma of the flower of another plant.

Oval. Broadly elliptic, the width over one-half the length. Figure 813.

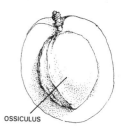

Figure 812

Ovary. The expanded basal portion of the pistil that contains the ovules. Figure 814.

Ovate. Egg-shaped in outline and attached at the broad end (applied to plane surfaces). Figure 815. (compare **ovoid**)

Ovoid. Egg-shaped (applied to three-dimensional structures). Figure 816.

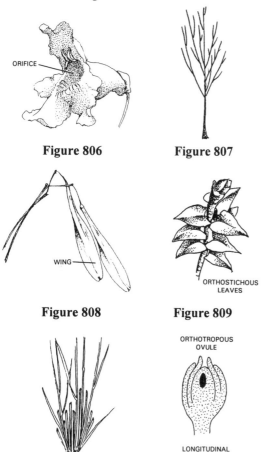

Figure 806   **Figure 807**

Figure 808   **Figure 809**

Figure 810   **Figure 811**

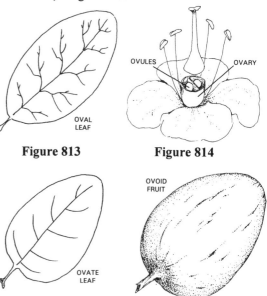

Figure 813   **Figure 814**

Figure 815   **Figure 816**

Ovulate. Of or pertaining to ovules; producing ovules.

Ovule. An immature seed; the megasporangium and surrounding integuments of a seed plant. Figure 817.

Osseous. Bony.

Ossiculus. The stone or pit of a drupe; a pyrene. Figure 812.

Ossified. Becoming bony.

Ovuliferous. Bearing ovules.

Oxylophyte. A plant growing on acidic soils.

Pachycarpous. With large, thick fruit.

Pachycladous. With thick branches.

Pachyphyllous. With thick leaves.

Pagina. A flat surface, such as the blade of a leaf. Figure 818.

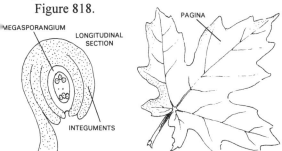

Figure 817   **Figure 818**

Palate. A raised appendage on the lower lip of a corolla which partially or completely closes the throat. Figure 819.

Pale. See **palea.**

Palea (pl. **paleae**). A chaffy scale or bract; the uppermost of the two bracts (lemma and palea) which subtend a grass floret, often partially surrounded by the lemma. Figure 820.

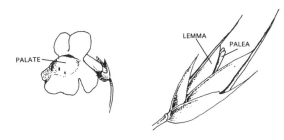

Figure 819   **Figure 820**

Paleaceous. Chaffy; with chaffy scales. Figure 821.

Paleola (pl. **paleolae**). A tiny palea; a lodicule. Figure 822.

Paleolate. With a lodicule. Figure 822.

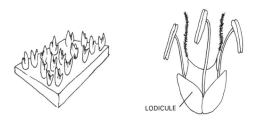

Figure 821   **Figure 822**

Paleomorphic. Lacking symmetry.

Palet. See **palea.**

Pallid. Pale.

Palmate. Lobed, veined, or divided from a common point, like the fingers of a hand. Figures 823 and 824. (compare **pinnate**)

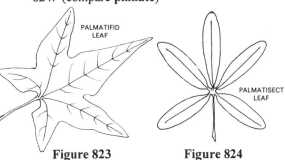

Figure 823   **Figure 824**

Palmate-pinnate. With the primary leaflets palmately arranged and the secondary leaflets pinnately arranged. Figure 825.

Palmatifid. Palmately cleft or lobed. Figure 823.

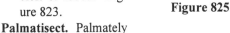

Figure 825

Palmatisect. Palmately divided. Figure 824.

Paludose. Growing in wet meadows or marshes. (same as **palustrine**)

Palustrine. See **paludose.**

Pampiniform. Tendril-like.

Pampinus. A tendril. Figure 826.

Pandurate. Fiddle-shaped. Figure 827.

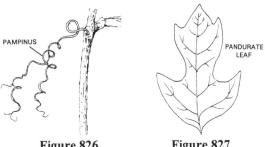

Figure 826   **Figure 827**

Panduriform. See **pandurate.**

Panicle. A branched, racemose inflorescence with flowers maturing from the bottom upwards. Figure 828.

Paniculate. Having flowers in panicles. Figure 828.

Paniculiform. An inflorescence with the general appearance, but not necessarily the structure, of a true panicle.

Panniform. See **pannose**.

Pannose. Covered with a short, dense, felt-like tomentum. Figure 829.

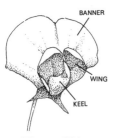

Figure 828   Figure 829

Papaveraceous. Resembling a poppy; belonging to the Papaveraceae.

Papilionaceous. Butterfly-like, as the irregular corolla of a pea, with a banner petal, two wing petals, and two fused keel petals. Figure 830.

Papilla (pl. **Papillae**). A short, rounded nipple-like bump or projection. Figure 831.

Papillary. Resembling papillae.

Papillate. Having papillae. Figure 831.

Figure 830   Figure 831

Papillose. Having minute papillae. Figure 832.

Papillose-hispid. With stiff hairs borne on swollen, nipple-like bases. Figure 833.

Pappiferous. Pappus-bearing. Figure 834.

Pappose. Pappus-bearing. Figure 834.

Figure 832

Pappus. The modified calyx of the Compositae

(Asteraceae), consisting of awns, scales, or bristles at the apex of the achene. Figure 834.

Papyraceous. Papery in texture and usually color.

Paracarp. An aborted ovary.

Paracarpium. See **paracarp**.

Parallelodromous. See **parallel-veined**.

Parallel-veined. With the main veins parallel to the leaf axis or to each other. Figure 835. (compare **net-veined**)

Paranema. See **paraphysis**.

Paraphyllum. See **stipule**.

Paraphysis (pl. **paraphyses**). A sterile filament occurring among the sporangia of some ferns. Figure 836.

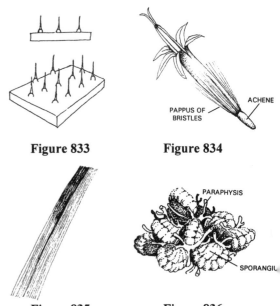

Figure 833   Figure 834

Figure 835   Figure 836

Parasite. An organism that obtains its food or water, at least partly, from a host organism. (compare **epiphyte**)

Parastemon. See **staminode**.

Parietal. Positioned along the edges or wall, rather than on the axis.

Parietal placentation. Ovules attached to the walls of the ovary. Figure 837.

Paripinnate. Even-pinnate; lacking a terminal leaflet. Figure 838.

Parted. Deeply cleft, usually more than half the distance to the base or midvein. Figure 839.

Parthenocarpy. Development of a fruit without

fertilization or seed production.

Parthenogenesis. Development from the egg without fertilization.

Parti-colored. Of different colors; variegated.

Patelliform. Shaped like a kneecap. Figure 840.

Patent. Spreading or expanded, as in widely spreading branches or broadly expanded petals. Figures 841 and 842.

Patulous. Open or spreading. Figures 841 and 842.

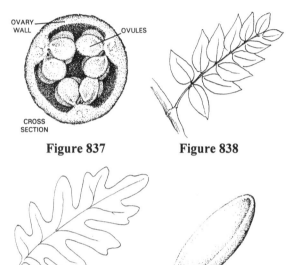

Figure 837   **Figure 838**

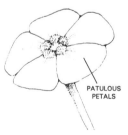

Figure 839   **Figure 840**

Figure 841   **Figure 842**

Pectinate. Comb-like; with close, regularly spaced divisions, appendages, or hairs, often in a single row, like the teeth of a comb. Figure 843.

Pedate. Palmately divided, with the lateral lobes 2-cleft. Figure 844.

Pedatifid. Pedately cleft. Figure 845.

Pedicel. The stalk of a single flower in an inflorescence, or of a grass spikelet. Figures 846 and 847.

Pedicellate. With a pedicel. Figures 846 and 847.

Pedicle. See **pedicel**.

Pediculus. See **pedicel**.

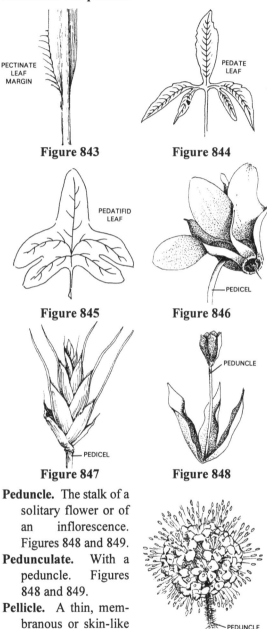

Figure 843   **Figure 844**

Figure 845   **Figure 846**

Figure 847   **Figure 848**

Peduncle. The stalk of a solitary flower or of an inflorescence. Figures 848 and 849.

Pedunculate. With a peduncle. Figures 848 and 849.

Pellicle. A thin, membranous or skin-like covering.

Pellucid. Transparent or translucent.

Figure 849

Peloria. Radial symmetry in flowers normally

bilaterally symmetrical.

Pelta. A scale or bract attached at the middle.

Peltafid. Peltate and divided into segments. Figure 850.

Peltate. Shield-shaped; a flat structure borne on a stalk attached to the lower surface rather than to the base or margin. Figure 851.

 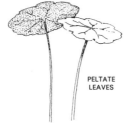

PELTATE LEAVES

Figure 850   **Figure 851**

Peltiform. See **peltate**.

Pencilled. See **penicillate**.

Pendent. See **pendulous**.

Pendulous. Hanging or drooping downward. Figure 852.

Penicil. A brush-like tuft of short hairs. Figure 853.

Penicillate. With a tuft of short hairs at the end, like a brush; with a penicil or penicils. Figure 853.

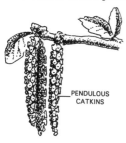 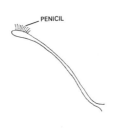

PENICIL

PENDULOUS CATKINS

Figure 852   **Figure 853**

Pennate. See **pinnate**.

Pennatifid. See **pinnatifid**.

Penni-parallel. See **penniveined**.

Penniveined. With parts arising parallel to one another from a main axis, like the veins of a feather. Figure 854.

Penta- (prefix).

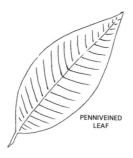

PENNIVEINED LEAF

Figure 854

Meaning five.

Pentacamerous. With five locules. Figure 855.

Pentacarpellary. With five carpels. Figure 856.

Pentacyclic. With five whorls.

Pentadactylous. Divided into five finger-like segments. Figure 857.

Pentadelphous. With the stamens arranged into five groups or clusters.

Pentagonal. Five-angled. Figure 858.

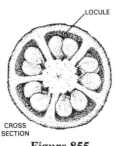 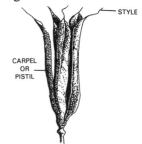

LOCULE

CROSS SECTION

STYLE

CARPEL OR PISTIL

Figure 855   **Figure 856**

PENTADACTYLOUS LEAF

Figure 857   **Figure 858**

Pentagynous. With five pistils or styles. Figure 856.

Pentamerous. With parts arranged in sets or multiples of five. Figure 859.

Pentandrous. With five stamens. Figure 859.

Pentapetalous. With five petals. Figure 859.

PETALS   SEPALS

STAMENS

Figure 859

Pentapterous. With five wings.

Pentasepalous. With five sepals. Figure 859.

Pentastichous. In five vertical rows or ranks.

Pepo. A fleshy, indehiscent, many-seeded fruit with a tough rind, as a melon or a cucumber. Figure 860.

Perennate. To renew, as when lateral shoots arise from a caudex.

Perennial. A plant that lives three or more years.

Perfect. With both male and female reproductive organs (stamens and pistils); bisexual. Figure 861.

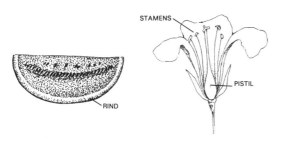

Figure 860   **Figure 861**

Perfoliate. A leaf with the margins entirely surrounding the stem, so that the stem appears to pass through the leaf. Figure 862.

Perforate. With holes or perforations. Figure 863.

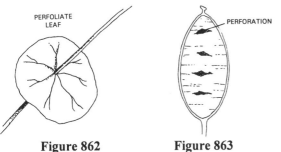

Figure 862   **Figure 863**

Perianth. The calyx and corolla of a flower, collectively, especially when they are similar in appearance. Figure 864.

Pericarp. The wall of the fruit. Figure 865.

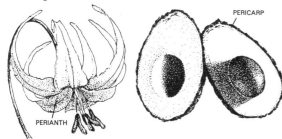

Figure 864   **Figure 865**

Periclinium. An involucre. Figure 866.

Peridroma. The rachis of a fern frond. Figure 867.

Perigynium (pl. **perigynia**). A scale-like bract enclosing the pistil in *Carex*. Figure 868.

Perigynous. With stamens, petals, and sepals borne on a calyx tube (hypanthium) surrounding, but not actually attached to, the superior ovary. Figure 869.

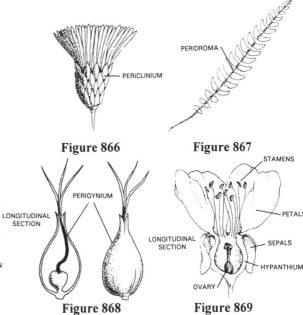

Figure 866   **Figure 867**

Figure 868   **Figure 869**

Peripheral. Outside of or external to; at the margin.

Peripterous. With a surrounding border or wing. Figure 870.

Perisperm. Food storage tissue in some seeds, arising from the nucellus.

Perispore. A membrane surrounding a spore, as in *Equisetum* spores. Figure 871.

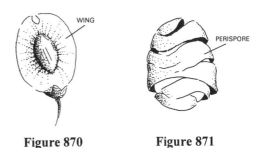

Figure 870   **Figure 871**

Pernicious. Harmful, destructive, or deadly in nature.

Persicicolor. Peach-colored.

Persistent. Remaining attached after similar parts are normally dropped, after the function has been completed.

Personate. Two-lipped, with the throat closed by a prominent projection (palate). Figure 872; masked.

Perspicuate. See **perspicuous**.

Perspicuous. Transparent.

Perula. See **perule**.

Perulate. With a perule. Figures 873 and 874.

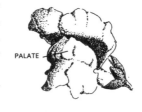

Figure 872

Perule. A leaf-bud scale. Figure 873; a sac or projection formed by the enlargement of two united sepals in some members of the Orchidaceae. Figure 874.

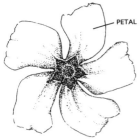

Figure 873   **Figure 874**

Petal. An individual segment or member of the corolla, usually colored or white. Figure 875.

Petalantherous. Of a stamen with a petaloid filament. Figure 876.

Petaliferous. Bearing petals. Figure 875.

Petaline. Of or pertaining to a petal; petaloid.

Petalode. An organ (usually a stamen) which resembles a petal. Figure 876.

Figure 875   **Figure 876**

Petalody. A condition in which various organs in a flower, such as stamens, become petals or become petaloid, as in some double flowers.

Petaloid. Petal-like in appearance.

Petalostemonous. With the staminal filaments fused to the corolla and the anthers free. Figure 877.

Petalous. With petals. Figure 875.

Petiolar. Pertaining to the petiole; growing from the petiole. Figure 878.

Petiolate. With a petiole. Figure 878.

Petiole. A leaf stalk. Figure 878.

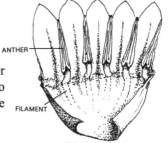
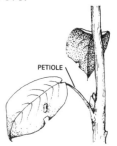

Figure 877   **Figure 878**

Petioled. See **petiolate**.

Petioliform. Resembling a petiole.

Petiolulate. With a petiolule. Figure 879.

Petiolule. The stalk of a leaflet of a compound leaf. Figure 879.

Phaenantherous. With stamens exserted from the corolla. Figure 880.

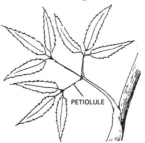
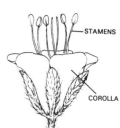

Figure 879   **Figure 880**

Phaenocarpous. With the ovary (fruit) free from the surrounding floral parts. Figure 881.

Phalange. Two or more stamens joined by their filaments. Figure 882.

Phalanx. See **Phalange**.

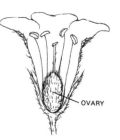

Figure 881

Phanerogam. A plant which produces seeds. (compare **cryptogam**)

Phellem. Cork.

Phloem. The food conducting tissue of vascular plants; bark. Figure 883.

Phoeniceous. Bright red.

Phoranthium. The receptacle of the flower head of the Compositae (Asteraceae). Figure 884.

Phragma (pl. **phragmata**). A septum or partition. Figure 885.

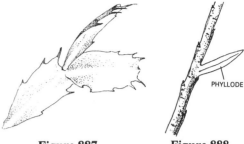

Figure 887   **Figure 888**

Phyllodium (pl. **phyllodia**). See **phyllode**.

Phylloid. Leaf-like.

Phyllome. A leaf. Figure 889.

Phyllopode. The dilated leaf base of an *Isoetes* leaf. Figure 890.

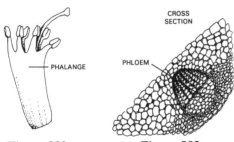

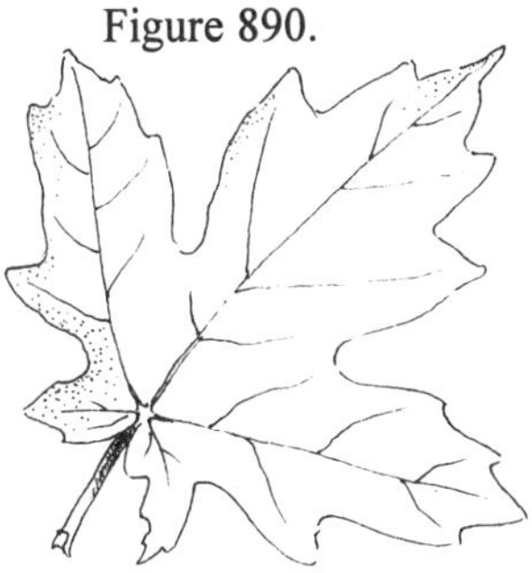

Figure 882   **Figure 883**

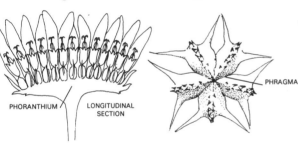

PHORANTHIUM   LONGITUDINAL SECTION

PHRAGMA

Figure 884   **Figure 885**

Phreatophyte. A plant with its root system typically in soil saturated with water.

Phyllary. An involucral bract of the Compositae (Asteraceae). Figure 886.

Phylloclad. See **phylloclade**.

Figure 886

Phylloclade. Part of a stem with the form and function of a leaf. Figure 887. (same as **cladophyll**)

Phyllode. An expanded, leaf-like petiole lacking a true leaf blade. Figure 888.

Phyllodic. Of or pertaining to a phyllode; with a phyllode. Figure 888.

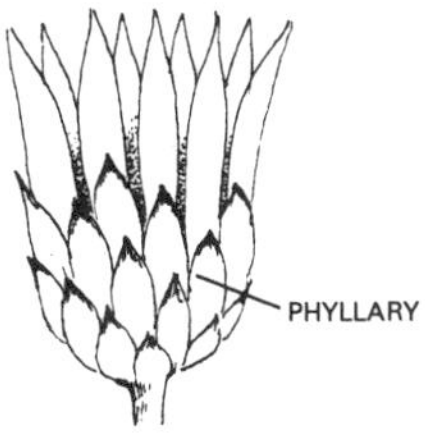

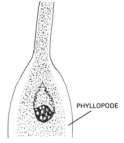

Figure 889   **Figure 890**

Phyllopodic. With the lowest leaves well developed, not reduced to scales. Figure 891. (compare **aphyllopodic**)

Phyllotaxis. See **phyllotaxy**.

Phyllotaxy. The arrangement of leaves on a stem.

Figure 891

When expressed as a fraction, the numerator indicates the number of turns around the stem, and the denominator indicates the number of internodes between two leaves in direct vertical alignment on the stem.

Phytomere. A section of a grass shoot including an internode, the leaf and a portion of the node at the top of the internode, and a portion of the node at the bottom of the internode. Figure 892.

Pileate. With a cap. Figure 893.

Piliferous. Tipped with a fine hair-like structure.

Figure 894.

Piliform. With the form of a hair.

Piloglandulose. With glandular hairs. Figure 895.

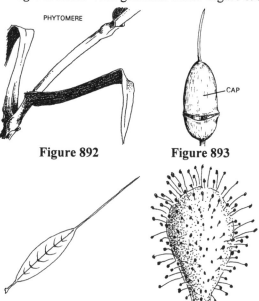

Figure 892　　　　**Figure 893**

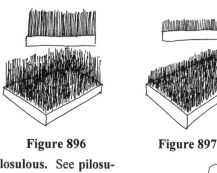

Figure 894　　　　**Figure 895**

Pilose. Bearing long, soft, straight hairs. Figure 896.

Pilosulose. Bearing minute, long, soft, straight hairs. Figure 897.

Figure 896　　　　**Figure 897**

Pilosulous. See **pilosu-lose**.

Pin. A heterostylic flower with a fairly long style and short stamens. Figure 898. (compare **thrum**)

Pinna (pl. **pinnae**). One of the primary divisions or leaflets of a

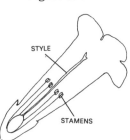

Figure 898

pinnate leaf. Figure 899.

Pinnate. Resembling a feather, as in a compound leaf with leaflets arranged on opposite sides of an elongated axis. Figure 899.

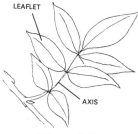

Figure 899

Pinnatifid. Pinnately cleft or lobed half the distance or more to the midrib, but not reaching the midrib. Figure 900.

Pinnatilobate. With pinnately arranged lobes. Figure 901.

Pinnation. Pinnate condition or development.

Pinnatipartite. Pinnately parted. Figure 902.

Pinnatisect. Pinnately cleft to the midrib. Figure 902.

Pinninerved. Pinnately veined. Figure 903.

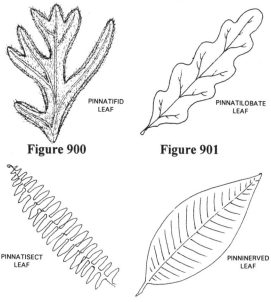

Figure 900　　　　**Figure 901**

Figure 902　　　　**Figure 903**

Pinnipalmate. Intermediate between pinnate and palmate, as in a leaf with the first pair of veins larger and more distinctive than the others. Figure 904.

Pinnule. The pinnate division of a pinna in a bipinnately compound leaf, or the ultimate divisions of a leaf which is more than twice pinnately compound. Figure 905.

Pip. A small seed of a fleshy fruit. Figure 906; one segment of a pineapple. Figure 907; a small blossom. Figure 908; short form of **pippin**.

Pippin. A seed or pip. Figure 906; an apple. Figure 909.

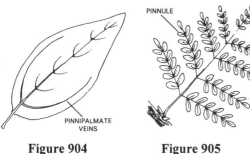

Figure 904

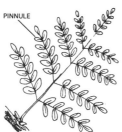

Figure 905

Figure 906

Figure 907

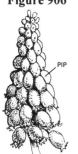

Figure 908

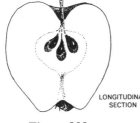

Figure 909

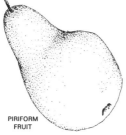

Figure 910

Piriform. Pear-shaped. Figure 910.

Pisaceous. Pea-green.

Pisiferous. Bearing peas.

Pisiform. Pea-shaped. Figure 911.

Pistil. The female reproductive organ of a flower, typically consisting of a stigma,

style, and ovary. Figure 912. (compare **gynoecium**)

Pistillate. Bearing a pistil or pistils, but lacking stamens. Figure 913. (same as **carpellate**; compare **staminate**)

Pistillode. A sterile, rudimentary pistil in an otherwise staminate flower. Figure 914.

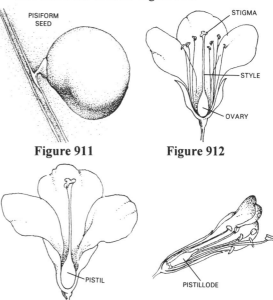

Figure 911

Figure 912

Figure 913

Figure 914

Pit. A small depression. Figure 915; the stony endocarp of a drupe, as in a peach or cherry. Figure 916; the small openings in the walls of tracheids and vessel elements which allow water to move from cell to cell. Figure 917.

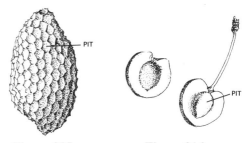

Figure 915

Figure 916

Pith. The spongy, parenchymatous central tissue in some stems and roots. Figure 918.

Pitted. With small pits or depressions. Figure 915.

Placenta (pl. **placentae**). The portion of the ovary bearing ovules. Figure 919.

Placentation. The arrangement or configuration of the placentae. (see **axile**, **basal**, **free central**, and **parietal** placentation)

Plait. A fold or pleat, as in some corollas. Figure 920.

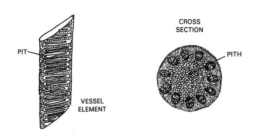

Figure 917   **Figure 918**

Figure 919   **Figure 920**

Plane. With a flat surface.

Plano-compressed. Compressed and flattened. Figure 921.

Plano-convex. Flat on one side and convex on the other. Figure 922.

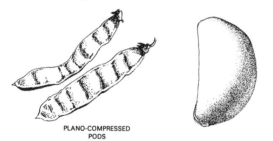

Figure 921   **Figure 922**

Pleiochasium. A cymose inflorescence with more than two branches from the main axis. Figure 923.

Pleiomery. The condition of having more than the usual number of floral whorls.

Pleiopetalous. With many petals.

Pleiosepalous. With many sepals.

Pleiospermous. With many seeds.

Plica (pl. **plicae**). A plait or fold. Figure 924.

Plicate. Plaited or folded, as a folding fan. Figure 924.

Pliestesial. Living several years before flowering and fruiting, and then dying, as in *Agave*.

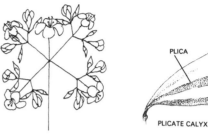

Figure 923   **Figure 924**

Plococarpium. A fruit consisting of follicles around an axis. Figure 925.

Plumbeous. Lead-colored.

Plumose. Feathery; with hairs or fine bristles on both sides of a main axis, as a plume. Figure 926.

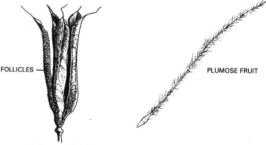

Figure 925   **Figure 926**

Plumule. The portion of the embryo above the point of attachment of the cotyledon(s) which gives rise to the shoot. Figure 927. (same as **epicotyl**)

Pluri- (prefix). Meaning many or several.

Pluricellular. Of many cells. Figure 928.

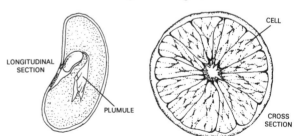

Figure 927   **Figure 928**

Pluricipital. With many heads, as in a highly branched caudex. Figure 929.

Plurilocular. See **pluricellular**.

Pluriovulate. With many ovules.

Pluriseriate. In many series or rows. Figure 930.

Pod. Any dry, dehiscent fruit, especially a legume or follicle. Figure 931.

Podocarp. A fruit borne on a gynophore. Figure 932.

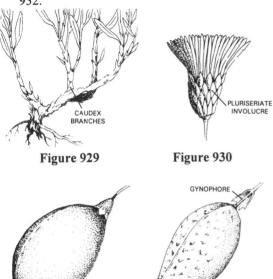

Figure 929

Figure 930

CAUDEX BRANCHES

PLURISERIATE INVOLUCRE

LEGUME

GYNOPHORE

Figure 931

Figure 932

Podogyne. See **carpopodium**.

Pollen. The mature microspores or developing male gametophytes of a seed plant, produced in the microsporangium of a gymnosperm or in the anther of an angiosperm. Figure 933.

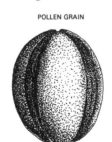

POLLEN GRAIN

Figure 933

Pollination. The transfer of pollen from the anther to the stigma.

Polliniferous. Bearing pollen.

Pollinium (pl. **pollinia**). A mass of waxy pollen grains transported as a unit in many members of the Orchidaceae and Asclepiadaceae. Figures 934 and 935.

Polsterform. Shaped like a low, compact mound. Figure 936.

Poly- (prefix). Meaning many.

Polyadelphous. Borne in several distinct groups, as the stamens of some flowers. Figure 937. (compare **monadelphous** and **diadelphous**)

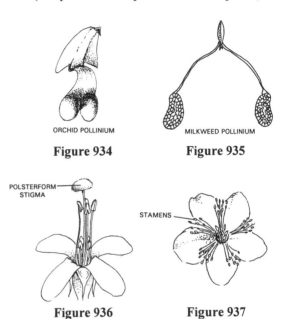

ORCHID POLLINIUM

MILKWEED POLLINIUM

Figure 934

Figure 935

POLSTERFORM STIGMA

STAMENS

Figure 936

Figure 937

Polyandrous. With many stamens (usually more than ten). Figure 938.

Polyanthous. With many flowers, especially when clustered together in an involucre. Figure 939.

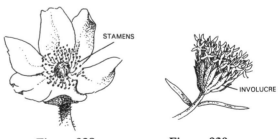

STAMENS

INVOLUCRE

Figure 938

Figure 939

Polycarpic. See **perennial**.

Polycarpous. With many carpels. Figure 940.

Polycephalous. With many flower heads. Figure 941.

Polychasium. A cymose inflorescence in which each axis produces more than two lateral axes.

Polychrome. Many colored.

Polycyclic. With many whorls.

Polygamo-dioecious. Mostly dioecious, but with some perfect flowers.

Polygamo-monoecious. Mostly monoecious, but with some perfect flowers.

Polygamous. With unisexual and bisexual flowers on the same plant.

Polygonal. Many-angled.

Polygynous. With many pistils or styles. Figure 940.

Polymerous. With many parts, as in a floral whorl with many members.

Polymorphic. Variable; with many forms.

Polymorphous. See **polymorphic**.

Polypetalous. A corolla of completely separate petals. Figure 943. (same as **apopetalous**; compare **gamopetalous** and **sympetalous**)

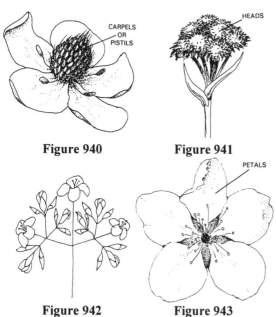

Figure 940   Figure 941

Figure 942   Figure 943

Polyploid. With three or more complete sets of chromosomes in each cell.

Polysepalous. A calyx of separate sepals. Figure 944. (compare **synsepalous** and **gamosepalous**)

Polystachyous. With many spikes, as in a grass with many ears or spikes.

Polystemonous. With many stamens (more than twice the number of petals or sepals). Figure

Figure 942.

945.

Polystichous. Arranged into several rows.

Polystigmous. With many stigmas.

Polystyllous. With many styles. Figure 946.

Polytrichous. Hairy. Figure 947.

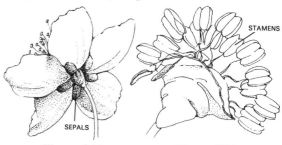

Figure 944   Figure 945

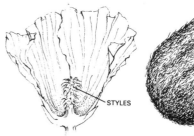

Figure 946   Figure 947

Pomaceous. Of or relating to a pome; pome-like.

Pome. A fleshy, indehiscent fruit derived from an inferior, compound ovary, consisting of a modified floral tube surrounding a core, as in an apple. Figure 948.

Porandrous. With anthers opening by pores. Figure 949.

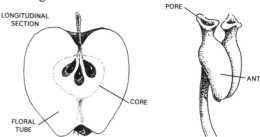

Figure 948   Figure 949

Pore. A small opening. Figure 950.

Poricidal. Opening by pores, as in a poppy capsule. Figure 950.

Porose. With pores. Figure 950.

Porrect. Extended forward; resembling a parrot

beak. Figure 951.

Posterior. At the back; on the side toward the axis, as the upper lip of a bilabiate corolla. Figure 952. (compare **anterior**)

Praemorse. Terminating abruptly, as if bitten off. Figure 953.

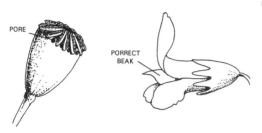

Figure 950   **Figure 951**

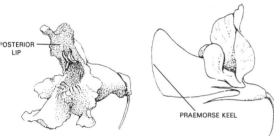

Figure 952   **Figure 953**

Precocious. Developing or appearing very early; with the flowers developing before the leaves.

Prehensile. Adapted for grasping, as in a tendril. Figure 954.

Prevernal. Pre-spring; flowering in early spring.

Prickle. A small, sharp outgrowth of the epidermis or bark. Figure 955. (compare **spine** and **thorn**)

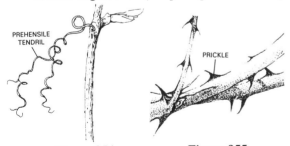

Figure 954   **Figure 955**

Primary. First, as the first division of a leaf which is more than once compound. Figure 956.

Primine. The outer integument layer of an ovule. Figure 957.

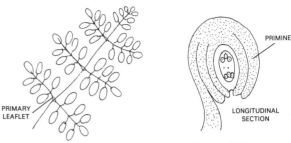

Figure 956   **Figure 957**

Primocane. The first-year, usually flowerless, cane (shoot) of *Rubus*. (compare **floricane**)

Prismatic. With sharp, definite angles and flat sides, like a prism. Figure 958; brilliant.

Process. An outgrowth or appendage. Figures 959 and 960.

Procumbent. Lying or trailing on the ground, but not rooting at the nodes. Figure 961.

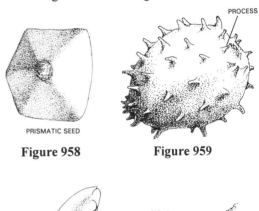

Figure 958   **Figure 959**

Figure 960   **Figure 961**

Projected. Extending outward. Figure 962.

Proliferous. Bearing plantlets or bulblets, usually from the leaves. Figure 963.

Prominent. Standing out from the surrounding surface, as raised veins on the surface of a leaf. Figure 964.

Propagule. A structure, such as a seed or spore, which gives rise to a new plant. Figure 965.

Prophyll. One of the paired bracteoles subtending

the flowers in some *Juncus* species. Figure 966.

Prophyllate. With prophylls.

Prophyllum. See **prophyll**.

Prop root. Adventitious roots arising from lower nodes and providing support to a stem. Figure 967.

Figure 962

Figure 963

Figure 964

Figure 965

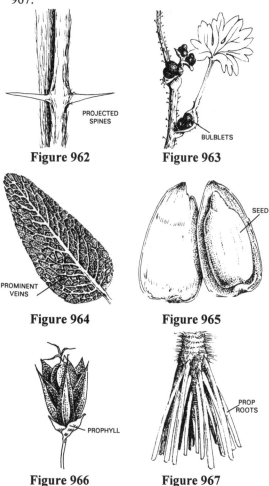

Figure 966

Figure 967

Prostrate. Lying flat on the ground. Figure 968.

Protandry (adj. **protandrous**). The anthers releasing pollen before the stigma is receptive.

Proterandry (adj. **proterandrous**). See **protandry**.

Figure 968

Proteranthy (adj. **proteranthous**). With the flowers developing before the leaves.

Proterogyny (adj. **proterogynous**). See **protogyny**.

Prothallium. See **prothallus**.

Prothallus (pl. **prothallia**). The small, usually flat, thallus-like growth germinating from a spore; the gametophyte generation in the alternation of generations. Figure 969.

Protogyny (adj. **protogynous**). The stigma receptive before the anthers release pollen.

Protostele. A stele with a solid core of vascular tissue, lacking a pith. Figure 970.

Protuberance. A rounded bulge, swelling, or projection. Figure 971.

Proximal. Toward the base, or the end of the organ by which it is attached. Figure 972. (compare **distal**)

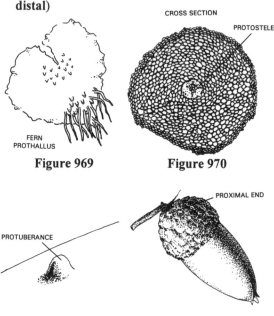

Figure 969

Figure 970

Figure 971

Figure 972

Pruinate. See **pruinose**.

Pruinose. With a waxy, powdery, usually whitish coating (bloom) on the surface; conspicuously glaucous, like a prune.

Pruniform. Plum-shaped. Figure 973.

Prurient. Causing itching.

Psammophyte. A plant growing in sand.

Pseudanthium. A compact inflorescence of many small flowers which simulates a single flower.

Figures 974 and 975.

Pseudo- (prefix). Meaning false.

Pseudobulb. A bulbous thickening on the stems of many epiphytic orchids. Figure 976.

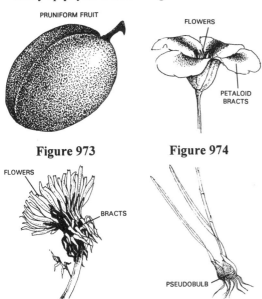

Figure 973

Figure 974

Figure 975

Figure 976

Pseudocarp. A fruit which develops from the receptacle rather than from the ovary, as in a pome. Figure 977.

Pseudofasciculate. Closely clustered, but not actually joined into a bundle. Figure 978.

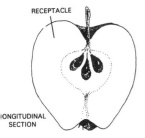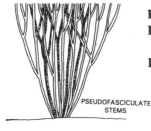

Figure 977

Figure 978

Pseudomonomerous. A structure which appears to be simple, though actually derived from the fusion of separate structures, as a pistil which appears to be composed of a single carpel, though actually composed of two or more carpels.

Pseudoscape. A false scape, where not all of the leaves are truly basal in origin though, superficially, they appear to be so. Figure 979.

Pseudoverticillate. Not actually whorled, but appearing so.

Pterocarpous. With winged fruits. Figure 980.

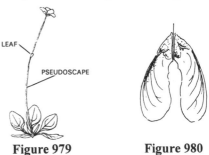

Figure 979

Figure 980

Pterocaulous. With winged stems. Figure 981.

Pterospermous. With winged seeds. Figure 982.

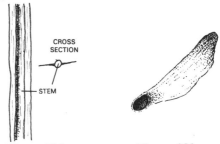

Figure 981

Figure 982

Pterygopous. With winged peduncles.

Puberulence. Fine, short hairs. Figure 983.

Puberulent. Minutely pubescent; with fine, short hairs. Figure 983.

Puberulous. See **puberulent**.

Pubescence. Hairiness; short, soft hairs. Figure 984.

Pubescent. Covered with short, soft hairs. Figure 984; bearing any kind of hairs.

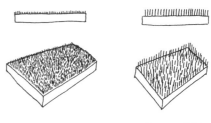

Figure 983

Figure 984

Pulveraceous. See **pulverulent**.

Pulverulent. Appearing dusty or powdery.

Pulvinate. Cushion-like or mat-like. Figure 985.

Pulviniform. See **pulvinate**.

Pulvinule. A small pulvinus at the base of a petiolule.

Pulvinus (pl. **pulvini**). A swelling or enlargement at the base of a petiole or petiolule. Figure 986.

Punctate. Dotted with pits or with translucent, sunken glands or with colored dots. Figure 987.

Puncticulate. Minutely punctate. Figure 988.

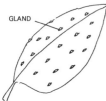

Figure 985

Figure 986

Figure 987

Figure 988

Punctiform. Reduced to a point.

Pungent. Tipped with a sharp, rigid point. Figure 989; with a sharp, acrid odor or taste.

Puniceous. Crimson colored.

Purpurescent. Becoming purplish.

Pustular. See **pustulose**.

Pustulate. See **pustulose**.

Pustule. Small blisterlike elevations. Figure 990.

Pustuliferous. See **pustulose**.

Pustulose. With small blisters or pustules, often at the base of a hair. Figure 990.

Putamen. The hard stony endocarp of some fruits.

Figure 989

Figure 991; a nut shell. Figure 992.

Pyramidal. Tetrahedral; pyramid-shaped. Figure 993.

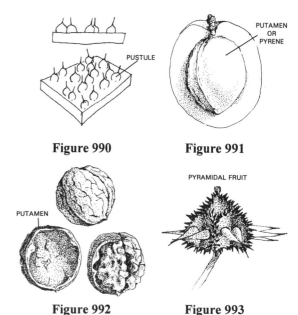

Figure 990

Figure 991

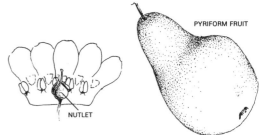

Figure 992

Figure 993

Pyrene. The stone or pit of a drupe or drupelet. Figure 991; a nutlet. Figure 994

Pyriform. Pear-shaped. Figure 995.

Figure 994

Figure 995

Pyxidate. With a pyxis. Figure 996.

Pyxidium. See **pyxis**.

Pyxis. A circumscissile capsule, the top coming off as a lid. Figure 996.

Quadrangular. Four-angled. Figure 997.

Quadrangulate. See **quadrangular**.

Figure 996

Quadrate. Square; rect-angular.

Quadri- (prefix). Meaning four.

Quadrifoliate. With four leaves or four leaflets. Figure 998.

Quadrilateral. With four sides. Figure 999.

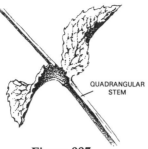

Figure 997

Quadripinnatifid. Four times pinnately cleft. Figure 1000.

Quilled. With tubular florets, especially in cases where the florets are typically ligulate, as in some members of the Compositae (Asteraceae). Figure 1001.

Quinary. In fives. Figure 1002.

Quinate. Five-parted. Figure 1003.

Quincuncial. With a five-ranked leaf arrangement. Figure 1004.

Quinquecostate. With five ribs. Figure 1005.

Figure 1004   **Figure 1005**

Quinquefarious. Arranged in five ranks. Figure 1006.

Quinquefoliate. With five leaves or five leaflets. Figure 1003.

Quinquejugate. Arranged in five pairs. Figure 1007.

Quinquelocular. With five cells or locules. Figure 1008.

Quinquenerved. With five main nerves. Figure 1009.

Quinquepartite. Divided into five parts. Figure 1003.

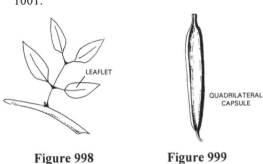

Figure 998   **Figure 999**

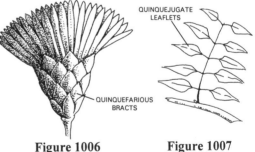

Figure 1000   **Figure 1001**   **Figure 1006**   **Figure 1007**

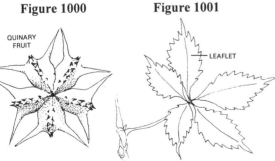

Figure 1002   **Figure 1003**   **Figure 1008**   **Figure 1009**

Raceme. An unbranch-
ed, elongated inflor-
escence with pedi-
cellate   flowers
maturing from the
bottom   upwards.
Figure 1010.

Racemiferous.   See
racemose.

Racemiform.   An in-
florescence with the

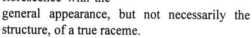

Figure 1010

general appearance, but not necessarily the
structure, of a true raceme.

Racemose. Having flowers in racemes. The term is
sometimes used in the same sense as **racemiform**. Figure 1010.

Rachilla. The axis of a grass or sedge spikelet.
Figure 1011; a small rachis.

Rachis. The main axis of a structure, such as a
compound leaf or an inflorescence. Figures 1012
and 1013.

Figure 1011   **Figure 1012**

Radial. With structures radiating from a central
point, as spokes on a wheel. Figure 1014; the
lateral spines of a cactus. Figure 1015.

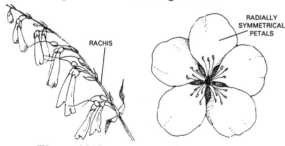

Figure 1013   **Figure 1014**

Radiant. See **radiate**.

Radiate. With parts spreading from a central point.
Figure 1016; in the Compositae (Asteraceae),

with some of the flowers of the involucrate head
ligulate (the petals united into a strap-like
corolla). Figure 1017.

Radical. Pertaining to the root; arising from, or
near, the roots.

Radicant. Rooting from the node of a prostrate stem
or from a leaf. Figure 1018.

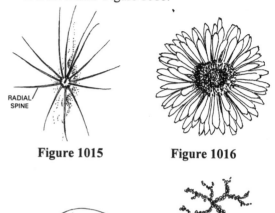

Figure 1015   **Figure 1016**

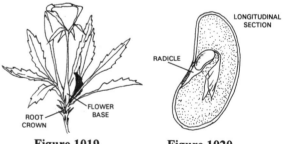

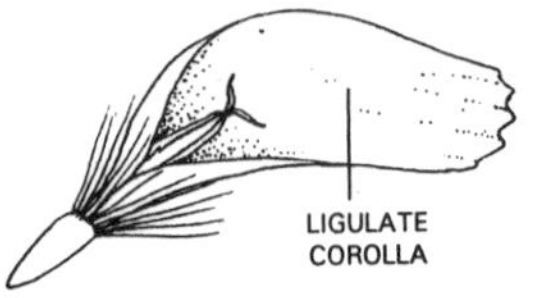

Figure 1017 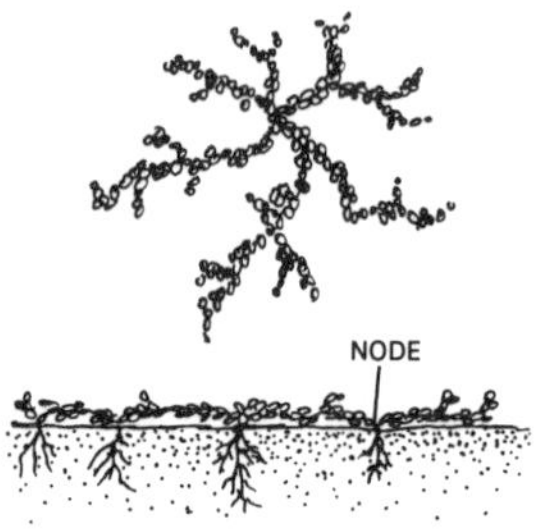   **Figure 1018**

Radicicolous. With the flower positioned directly
upon the root crown. Figure 1019.

Radicle. The part of the plant embryo which will
develop into the primary root. Figure 1020.

Figure 1019   **Figure 1020**

Ramal. See **rameal**.

Rameal. Pertaining to the branches.

Ramentaceous. Having ramentum. Figure 1021.

Ramentum. The flattened, scaly outgrowths on the
epidermis of the stem and leaves of some ferns.
Figure 1021.

Ramification. The arrangement of branching parts.

Ramiform. Branch-like in form; branched.

Ramose. With many branches; branching. Figure 1022.

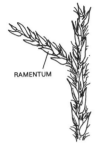

Figure 1021

Figure 1022

Ramous. See **ramose**.

Ramulose. See **ramose**.

Range. The area of distribution of a plant.

Rank. A vertical row, as in a plant with 2-ranked leaves arranged into two rows. Figure 1023.

Ranked. Arranged into vertical rows. Figure 1023.

Figure 1023

Raphal. Of or pertaining to the raphe.

Raphe (pl. **raphae**). A ridge on the seed formed by the portion of the funiculus fused to the seed coat. Figure 1024.

Rapiformis. Turnip-shaped. Figure 1025.

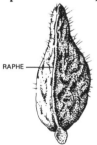

Figure 1024

Figure 1025

Ratoon. A shoot arising from the root of a plant that has been cut down. Figure 1026.

Ray. The strap-like portion of a ligulate flower (or the ligulate flower itself) in the Compositae (Asteraceae). Figure 1027; a branch of an umbel. Figure 1028.

Ray flower. A ligulate flower of the Compositae (Asteraceae). Figure 1027. (compare **disk flower**)

Receptacle. The portion of the pedicel upon which the flower parts are borne. Figure 1029; in the Compositae (Asteraceae), the part of the peduncle where the flowers of the head are borne. Figure 1030.

Reclinate. Bent abruptly downward. Figure 1031.

Reclining. Bending or curving downward. Figure 1031; lying upon something and being supported by it.

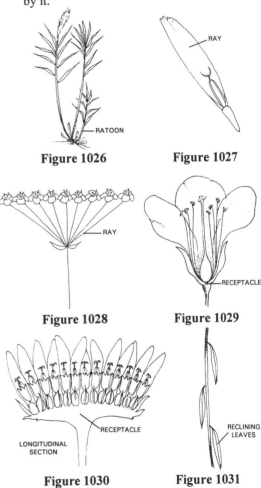

Figure 1026   **Figure 1027**

Figure 1028   **Figure 1029**

Figure 1030   **Figure 1031**

Recumbent. Leaning or resting on the ground; prostrate. Figure 1032.

Recurved. Curved backward. Figure 1033.

Reduced. Diminished in size.

Reduplicate. Valvate with the edges reflexed.

Figure 1034.

Reflexed. Bent backward or downward. Figure 1035.

Refoliate. To produce leaves again, as after rain, wind, or disease.

Refracted. Bent backward from the base. Figure 1035.

Regma (pl. **regmata**). A dry fruit of three or more carpels which separate at maturity. Figure 1036; a type of schizocarp.

Regular. Radially symmetrical; said of a flower in which all parts are similar in size and arrangement on the receptacle. Figure 1037. (compare **irregular**, and see **actinomorphic**)

Remote. Distantly spaced.

Reniform. Kidney-shaped. Figure 1038.

Repand. With a slightly wavy or weakly sinuate margin; undulate. Figure 1039.

Repent. Prostrate; creeping. Figure 1040.

Replicate. Folded backward. Figure 1035.

Replum. Partition or septum between the two valves or compartments of silicles or siliques in the Cruciferae (Brassicaceae). Figure 1041.

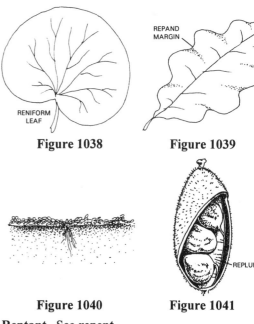

Figure 1038   **Figure 1039**

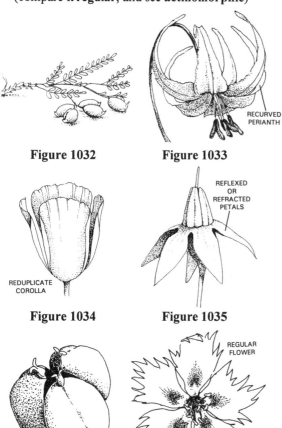

Figure 1032   **Figure 1033**

Figure 1034   **Figure 1035**

Figure 1036   **Figure 1037**

Figure 1040   **Figure 1041**

Reptant. See **repent**.

Resiniferous. See **resinous**.

Resinous. Bearing resin and often, therefore, sticky.

Resupinate. Upside down due to twisting of the pedicel, as the flowers of some orchids. Figure 1042.

Reticulate. In the form of a network. Figure 1043; net-veined. Figure 1044.

Regularly. Evenly or uniformly.

Relict. A plant which has survived from an earlier flora or from a past geologic epoch.

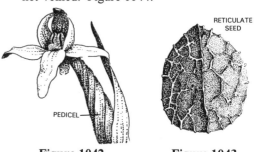

Figure 1042   **Figure 1043**

Reticulation. See **reticulum**.

Reticulum (pl. **reticula**). A network of veins or fibers. Figures 1043 and 1044.

Retinaculum. The sticky gland attached to the pollinia of the Orchidaceae. Figure 1045.

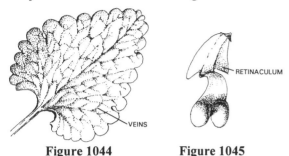

Figure 1044   Figure 1045

Retrocurved. See **recurved**.

Retroflexed. See **reflexed**.

Retrorse (Adv. **retrorsely**). Directed downward or backward. Figure 1046. (compare **antrorse**)

Retuse. With a shallow notch in a round or blunt apex. Figure 1047.

Revolute. With the margins rolled backward toward the underside. Figure 1048. (compare **involute**)

Rhabdocarpous. With long rod-shaped fruits. Figure 1049.

Rhachilla. See **rachilla**.

Rhachis. See **rachis**.

Rhaphe. See **raphe**.

Rhipidium. A flattened, fan-shaped cyme. Figure 1050.

Rhizanthous. With the flowers arising so close to the ground that they appear to be arising from the root. Figure 1051.

Rhizocarpic. With the roots living for several to many years and the stems dying each year.

Rhizocarpous. See **rhizocarpic**.

Rhizogenic. Root producing.

Rhizoid. A root-like structure lacking conductive tissues (xylem and phloem).

Rhizomatous. Rhizome-like; with rhizomes. Figure 1052.

Rhizome. A horizontal underground stem; rootstock. Figure 1052.

Rhizomorphous. Root-like in appearance.

Rhizophyllous. With roots arising from the leaves. Figure 1053.

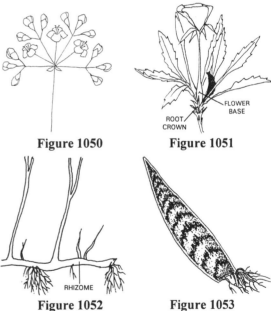

Figure 1050   Figure 1051

Figure 1052   Figure 1053

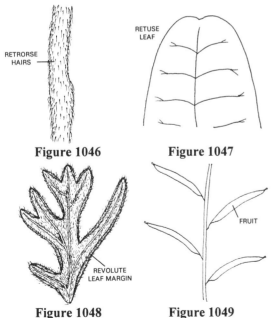

Figure 1046   Figure 1047

Figure 1048   Figure 1049

Rhizotaxis. See **rhizotaxy**.

Rhizotaxy. The type of arrangement of roots on a plant.

Rhombic. Diamond-shaped. Figure 1054.

Rhomboid. See **rhomboidal**.

Rhomboidal. Quadrangular, nearly rhombic, with obtuse lateral angles. Figure 1055.

Rib. A main longitudinal vein in a structure,

particularly if raised above the surrounding surface, as in some leaves and other organs. Figures 1056 and 1057.

Ribbed. With prominent ribs or veins. Figures 1056 and 1057.

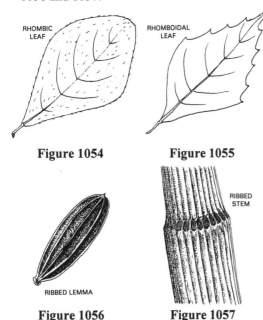

Figure 1054   **Figure 1055**

Figure 1056   **Figure 1057**

Rictus. The mouth of a bilabiate corolla. Figure 1058.

Rigescent. Becoming rigid.

Rigid. Stiff and inflexible.

Rim. A projecting edge or flange. Figure 1059.

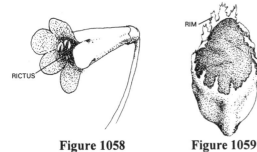

Figure 1058   **Figure 1059**

Rimose. With fissures or cracks, as in the bark of some trees. Figure 1060.

Rimous. See **rimose**.

Rind. A thick outer covering, as in the tough outer layer of a pepo. Figure 1061.

Ringent. Gaping; with widely spreading lips, as in some corollas. Figure 1062.

Figure 1060   **Figure 1061**

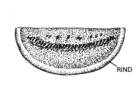

Riparian. Growing along the banks of streams, springs, or seeps.

Riparious. See **riparian**.

Ripe. Fully developed and mature.

Rivulose. With meandering channels or marked with sinuous lines resembling a rivulet. Figure 1063.

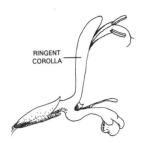

Figure 1062

Root. That portion of the plant axis lacking nodes and leaves and usually found below ground. Figure 1064.

Rootlet. A small root. Figure 1064.

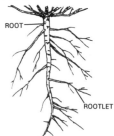

Figure 1063   **Figure 1064**

Rootstock. See **rhizome**.

Roridulate. With a covering of waxy platelets, appearing moist.

Roseate. Tinged with red; rosy.

Rosette. A dense radiating cluster of leaves

Figure 1065

(or other organs), usually at or near ground level. Figure 1065.

Rostellate. With a tiny, short, stout, terminal beak. Figure 1066.

Rostellum. A small beak. Figure 1066; an extension from the upper edge of the stigma in orchids. Figure 1067.

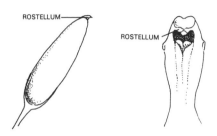

Figure 1066   **Figure 1067**

Rostrate. With a short, stout, terminal beak. Figure 1068.

Rostrum. A beak-like structure. Figure 1068.

Rosulate. With the leaves arranged in basal rosettes, the stem very short or lacking. Figure 1069.

Rotate. Disc-shaped; flat and circular, as a sympetalous corolla with widely spreading lobes and little or no tube. Figure 1070.

Rotund. Round or rounded in outline. Figure 1071.

Rotundifolious. With round leaves. Figure 1071.

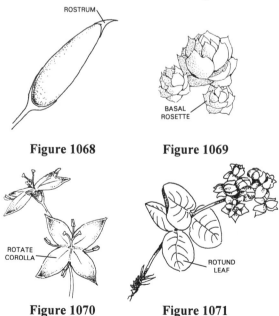

Figure 1068   **Figure 1069**

Figure 1070   **Figure 1071**

Rubescent. Becoming red or reddish.

Rubiginose. See **rubiginous**.

Rubiginous. Rust-colored.

Ruderal. Growing in disturbed habitats; weedy.

Rudimentary. Imperfectly developed; vestigial. Figure 1072.

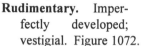

Figure 1072

Rufescent. See **rubescent**.

Rufous. Reddish-brown.

Rufus. See **rufous**.

Ruga (pl. **rugae**). A fold or wrinkle. Figure 1073.

Rugate. See **rugose**.

Rugose. Wrinkled. Figure 1073.

Rugulose. Slightly wrinkled. Figure 1074.

Ruminate. Roughly wrinkled, as if chewed. Figure 1075.

Runcinate. Sharply pinnatifid or cleft, the segments directed downward. Figure 1076.

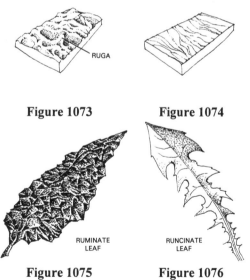

Figure 1073   **Figure 1074**

Figure 1075   **Figure 1076**

Runner. A slender stolon or prostrate stem rooting at the nodes or at the tip. Figure 1077.

Ruptile. Dehiscing irregularly. Figure 1078.

Rush-like. Grass-like in appearance, with inconspicuous flowers. Figure 1079.

Sac. A bag-shaped compartment, as the cavity of an anther or the lower lip of some corollas. Figures 1080 and 1081.

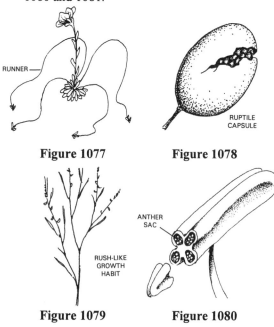

Figure 1077   **Figure 1078**

Figure 1079   **Figure 1080**

Saccate. With a sac, or in the shape of a sac; bag-shaped. Figure 1081.

Sacciform. See **saccate**.

Sacculate. With a saccule, or in the shape of a saccule. Figure 1082.

Saccule. A very small sac or cavity. Figure 1082.

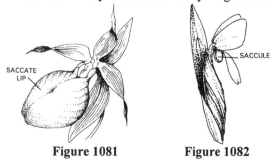

Figure 1081   **Figure 1082**

Sacculus (pl. **sacculi**). See **saccule**.

Sagittate. Arrowhead-shaped, with the basal lobes directed downward. Figure 1083. (compare **hastate**)

Sagittiform. See **sagittate**.

Salient. Projecting outward. Figure 1084.

Salverform. With a slender tube and an abruptly spreading, flattened limb. Figure 1085.

Samara. A dry, indehiscent, winged fruit. Figure 1086.

Samaroid. Samara-like.

Sanguine. Blood red.

Sanguineous. Blood red.

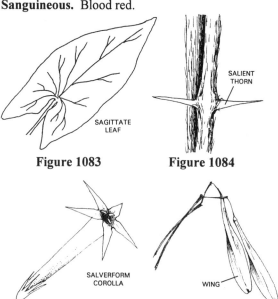

Figure 1083   **Figure 1084**

Figure 1085   **Figure 1086**

Sap. The juice of a plant; the fluids circulated throughout a plant.

Sapid. With an agreeable taste.

Saponaceous. Soapy, as in a substance or object slippery to the touch.

Sapor. The flavor or taste of a plant or plant substance.

Saprobe (Adj. **saprobic**). See **saprophyte**.

Saprophyte (Adj. **saprophytic**). A plant living on dead organic matter, lacking chlorophyll. (compare **parasite**)

Sapwood. The outer, newer, usually somewhat lighter, wood of a woody stem; the wood that is actively transporting water; alburnum. Figure 1087.

Sarcocarp. The fleshy portion (mesocarp)

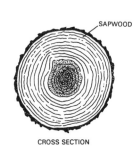

Figure 1087

of a fleshy fruit. Figure 1088.

Sarcocaulous. With fleshy stems. Figure 1089.

Sarcous. Fleshy. Figure 1090.

Sarment. A long, slender runner. Figure 1091.

Sarmentose. With long, slender runners. Figure 1091.

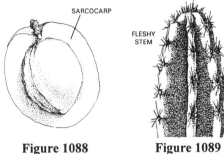

Figure 1088   **Figure 1089**

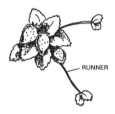

Figure 1090   **Figure 1091**

Scaberulent. See **scaberulose**.

Scaberulose. Slightly rough to the touch, due to the structure of the epidermal cells, or to the presence of short stiff hairs. Figure 1092.

Scaberulous. See **scaberulose**.

Scabrellate. See **scaberulose**.

Scabrid. Roughened.

Scabridulous. Minutely roughened.

Scabrous. Rough to the touch, due to the structure of the epidermal cells, or to the presence of short stiff hairs. Figure 1093.

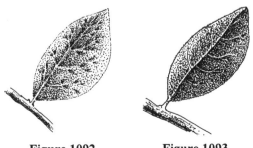

Figure 1092   **Figure 1093**

Scalariform. Ladder-like. Figure 1094.

Scale. Any thin, flat, scarious structure. Figures 1095 and 1096.

Scandent. Climbing. Figure 1097.

Scape. A leafless peduncle arising from ground level (usually from a basal rosette) in acaulescent plants. Figure 1098.

Scaphoid. Boat-shaped. Figure 1099.

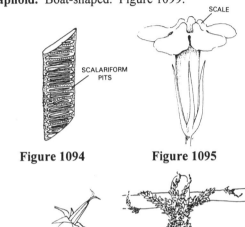

Figure 1094   **Figure 1095**

Figure 1096   **Figure 1097**

Figure 1098   **Figure 1099**

Scapiflorous. See **scapose**.

Scapiform. Scape-like but not entirely leafless. Figure 1100.

Scapose. With flowers borne on a scape. Figure 1098; scape-like. Figure 1100.

Scar. The mark left on a seed after detachment from the placenta. Figure 1101; the mark left on a stem after leaf abscission. Figure 1102.

Scarious. Thin, dry, and membranous in texture, not

green. Figure 1103.

Scattered. Irregularly, and usually sparsely, arranged. Figure 1104.

Schizocarp. A dry, indehiscent fruit which splits into separate one-seeded segments (carpels) at maturity. Figure 1105.

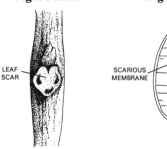

Figure 1100

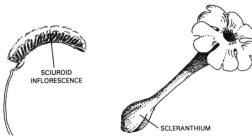

Figure 1101

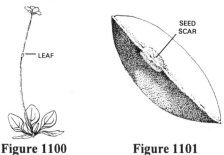

Figure 1102

Figure 1103

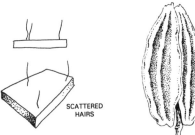

Figure 1104

Figure 1105

Figure 1106

Schizogenous. Formed by the splitting or separation of tissue.

Schizopetalous. With cut petals. Figure 1106.

Scissile. Splitting easily.

Sciuroid. Shaped like the tail of a squirrel, as in some grass in-florescences. Figure

1107.

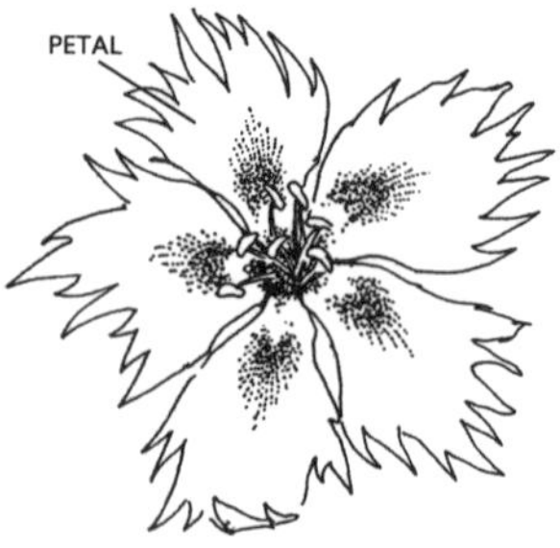

Scleranthium. An achene enclosed within a hardened calyx tube. Figure 1108.

Scleroid. See **sclerotic**.

Sclerophyll. A stiff, firm leaf which retains its stiffness even when wilted.

Figure 1107

Figure 1108

Sclerophyllous. With stiff, firm leaves; with sclerophylls.

Sclerosis. A hardening or thickening of tissue due to lignification.

Sclerotic. Hardened or thickened.

Sclerous. See **sclerotic**.

Scobiform. Sawdust-like in appearance.

Scobina. The zigzag rachilla of some grass spikelets. Figure 1109.

Scobinate. With a roughened surface, as though rasped. Figure 1110.

Figure 1109

Figure 1110

Scorpioid. Shaped like a scorpion's tail, as in some coiled cymes. Figure 1111; a determinate inflorescence with a zigzag rachis. Figure 1112.

Scrobiculate. Pitted or furrowed. Figures 1113 and 1114.

Scrotiform. Scrotum-like in appearance. Figure 1115; pouch-shaped.

Scurf. Small bran-like scales. Figure 1116.

Scurfy. Covered with small, bran-like scales. Figure 1116.

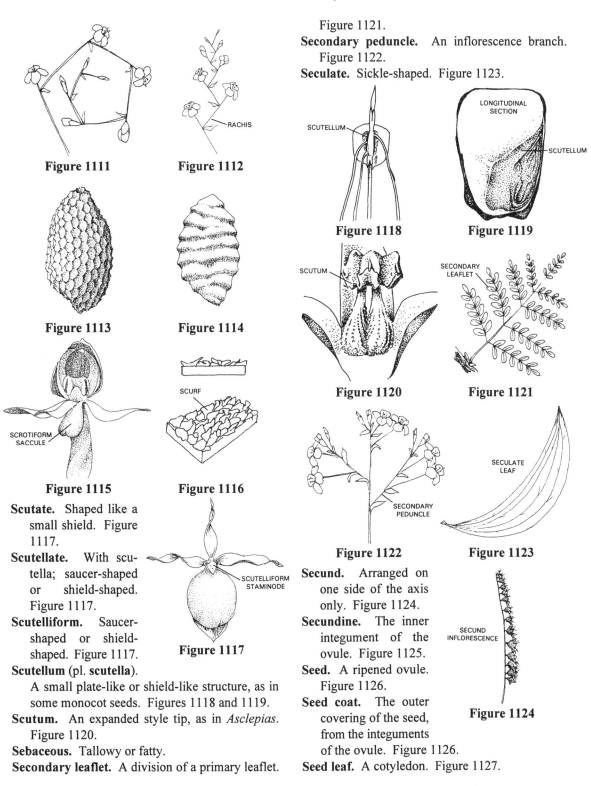

Figure 1111

Figure 1112

RACHIS

Figure 1113

Figure 1114

SCROTIFORM
SACCULE

Figure 1115

SCURF

Figure 1116

Scutate. Shaped like a small shield. Figure 1117.

Scutellate. With scutella; saucer-shaped or shield-shaped. Figure 1117.

Scutelliform. Saucer-shaped or shield-shaped. Figure 1117.

SCUTELLIFORM
STAMINODE

Figure 1117

Scutellum (pl. **scutella**). A small plate-like or shield-like structure, as in some monocot seeds. Figures 1118 and 1119.

Scutum. An expanded style tip, as in *Asclepias*. Figure 1120.

Sebaceous. Tallowy or fatty.

Secondary leaflet. A division of a primary leaflet.

Figure 1121.

Secondary peduncle. An inflorescence branch. Figure 1122.

Seculate. Sickle-shaped. Figure 1123.

SCUTELLUM

Figure 1118

LONGITUDINAL
SECTION

SCUTELLUM

Figure 1119

SCUTUM

Figure 1120

SECONDARY
LEAFLET

Figure 1121

SECULATE
LEAF

SECONDARY
PEDUNCLE

Figure 1122

Figure 1123

Secund. Arranged on one side of the axis only. Figure 1124.

Secundine. The inner integument of the ovule. Figure 1125.

Seed. A ripened ovule. Figure 1126.

Seed coat. The outer covering of the seed, from the integuments of the ovule. Figure 1126.

Seed leaf. A cotyledon. Figure 1127.

SECUND
INFLORESCENCE

Figure 1124

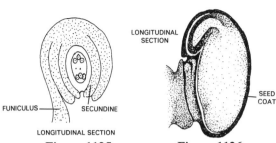

Figure 1125   **Figure 1126**

Seed stalk. The funiculus. Figures 1125 and 1128.

Segment. A section or division of an organ. Figure 1129.

Sejugous. With six pairs of leaflets. Figure 1130.

Figure 1127   **Figure 1128**

Figure 1129   **Figure 1130**

Seleniferous. Bearing selenium.

Selenophyte. A plant that grows on seleniferous soils and takes up selenium from these soils.

Self-pollination. Transfer of pollen from the anthers to the stigma of the same flower or to the stigma of another flower on the same plant.

Semen. A seed. Figure 1131.

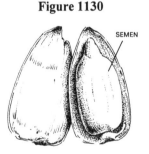

Figure 1131

Semi- (prefix). Half; partly or almost.

Semicarpous. With ovaries of carpels partly fused, the styles and stigmas separate. Figure 1132.

Seminiferous. Seed-bearing.

Semitropical. See **subtropical**.

Semperflorous. Flowering throughout the year.

Sensitive. Responsive to touch.

Sepal. A segment of the calyx. Figure 1133.

Sepaloid. Sepal-like in color and texture.

Septate. Divided by one or more partitions (septa). Figure 1134.

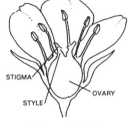

Figure 1132

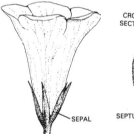

Figure 1133   **Figure 1134**

Septenate. With parts in sevens. Figure 1135.

Septicidal. Dehiscing through the septa and between the locules. Figure 1136. (compare **loculicidal** and **poricidal**)

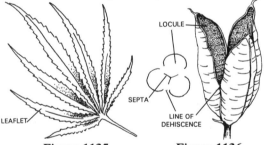

Figure 1135   **Figure 1136**

Septiferous. With a septum or septa. Figure 1137.

Septifolious. With seven leaves or seven leaflets. Figure 1135.

Septifragal. With valves separating from the septa at dehiscence. Figure 1138.

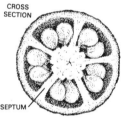
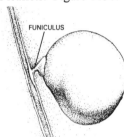

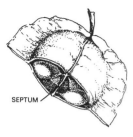

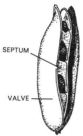

Figure 1137   **Figure 1138**

Septum (pl. **septa**). A partition, as the partitions separating the locules of an ovary. Figure 1134.

Seriate. Arranged in rows or series. Figure 1139.

Sericeous. Silky, with long, soft, slender, somewhat appressed hairs. Figure 1140.

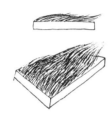

Figure 1139   **Figure 1140**

Serotinal. See **serot-inous**.

Serotinous. Late; late in flowering or leafing; with flowers developing after the leaves are fully developed.

Serra (pl. **serrae**). A tooth of a serrate leaf. Figure 1141.

Serrate. Saw-like; toothed along the margin, the sharp teeth pointing forward. Figure 1141.

Figure 1141

Serration. A serrated margin. Figure 1141; one of the teeth along a serrated margin; a serrated condition.

Serriform. See **serrate**.

Serrulate. Toothed along the margin with minute, sharp, forward-pointing teeth. Figure 1142.

Serrulation. A serrulate margin. Figure 1142; one of the teeth along a serrulate margin; a serrulate

condition.

Sessile. Attached directly, without a supporting stalk, as a leaf without a petiole. Figure 1143.

Seta (pl. **setae**). A bristle. Figure 1144.

Setaceous. Bristle-like; with bristles. Figure 1144.

Setiferous. Bristle-bearing. Figure 1144.

Setiform. Bristle-like.

Setose. Covered with bristles. Figure 1144.

Setule. A small bristle. Figure 1145.

Setulose. Covered with minute bristles. Figure 1145.

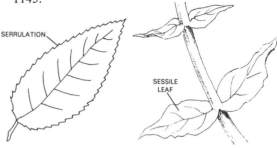

Figure 1142   **Figure 1143**

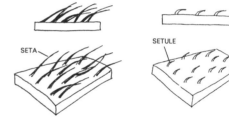

Figure 1144   **Figure 1145**

Sheath. The portion of an organ which surrounds, at least partly, another organ, as the leaf base of a grass surrounds the stem. Figure 1146.

Sheathing. Forming a sheath, as the leaf base of a grass forms a sheath as it surrounds the stem. Figure 1146.

Shield. The staminode of *Cypripedium*. Figure 1147; the outermost portion of a cone scale of a conifer cone. Figure 1148.

Shoot. A young stem or branch. Figure 1149.

Shrub. A woody plant,

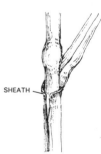

Figure 1146

with several stems, that is shorter than a typical tree. Figure 1150.

Sigmoid. S-shaped; doubly curved, like the letter S. Figure 1151.

Siliceous. Relating to, or containing silica.

Silicious. See **siliceous**.

Silicle. A dry, dehiscent fruit of the Cruciferae (Brassicaceae), typically less than twice as long as wide, with two valves separating from the persistent placentae and septum (replum). Figures 1152 and 1153.

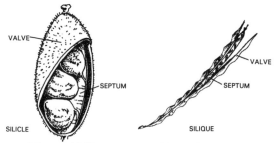

Figure 1153　　　　　　**Figure 1154**

Silky. Silk-like in appearance or texture; sericeous.

Simple. Undivided, as a leaf blade which is not separated into leaflets (though the blade may be deeply lobed or cleft). Figures 1156 and 1157; single, as a pistil composed of only one carpel. Figure 1158; unbranched, as a stem or hair. Figure 1159.

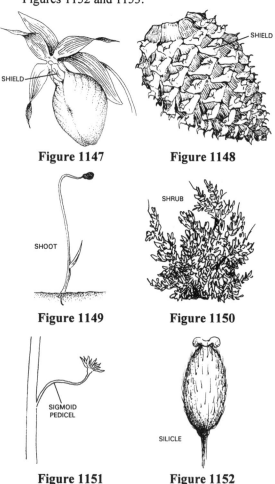

Figure 1147　　　　　　**Figure 1148**

Figure 1149　　　　　　**Figure 1150**

Figure 1151　　　　　　**Figure 1152**

Silique. A dry, dehiscent fruit of the Cruciferae (Brassicaceae), typically more than twice as long as wide, with two valves separating from the persistent placentae and septum (replum). Figure 1154.

Silk. The hair-like styles in maize. Figure 1155.

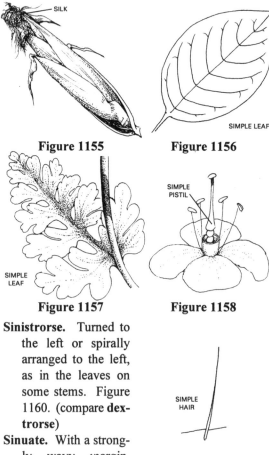

Figure 1155　　　　　　**Figure 1156**

Figure 1157　　　　　　**Figure 1158**

Sinistrorse. Turned to the left or spirally arranged to the left, as in the leaves on some stems. Figure 1160. (compare **dextrorse**)

Sinuate. With a strongly wavy margin. Figure 1161.

Figure 1159

(compare **undulate** or **repand**)

Sinuous. Of a wavy or serpentine form. Figure 1162.

Sinus. The cleft, depression, or recess between two lobes of an expanded organ such as a leaf or petal. Figure 1163.

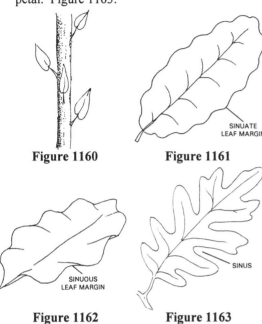

Figure 1160   **Figure 1161**

Figure 1162   **Figure 1163**

Smooth. With an even surface; not rough to the touch.

Sobol. Elongated caudex branches; a shoot arising from the base of a stem or from the rhizome. Figure 1164.

Sobole. See **sobol**.

Soboliferous. Of or pertaining to sobols; bearing sobols. Figure 1164.

Socket. A hollowed area formed to receive an articulating part. Figure 1165.

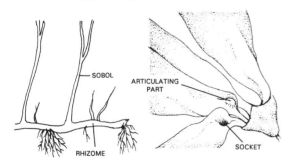

Figure 1164   **Figure 1165**

Sole. That end of the carpel most distant from the apex. Figure 1166.

Solitary. Occurring singly and not borne in a cluster or group. Figure 1167.

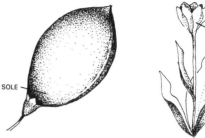

Figure 1166   **Figure 1167**

Somatic. Pertaining to or of the body, as all of the cells of a plant except the egg and sperm.

Sordid. Of a dull, dingy, or muddy color.

Sorose. See **sorosis**.

Sorosis. A fleshy multiple fruit arising from many flowers, as in the mulberry or pineapple. Figure 1168.

Sorus (pl. **sori**). A cluster of sporangia on the surface of a fern leaf. Figure 1169.

Figure 1168   **Figure 1169**

Spadiceous. Spadix bearing. Figure 1170; spadix-like.

Spadix. A spike with small flowers crowded on a thickened axis. Figure 1170.

Spananthus. With few flowers.

Spathaceous. Spathe bearing. Figure 1170; spathe-like.

Spathe. A large bract or pair of bracts subtending and often enclosing an inflorescence. Figure 1170.

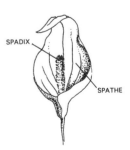

Figure 1170

Spathella. An archaic term referring most commonly to the lemma of a grass flower, but occasionally referring to a glume of a grass spikelet. Figures 1171 and 1172.

Spathellula. The palea of a grass flower. Figures 1171 and 1172.

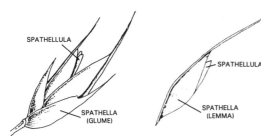

Figure 1171   **Figure 1172**

Spathiform. With the form of a spathe.

Spathulate. See **spatulate.**

Spatulate. Like a spatula in shape, with a rounded blade above gradually tapering to the base. Figure 1173.

Speiranthy. The condition of having twisted flowers. Figure 1174.

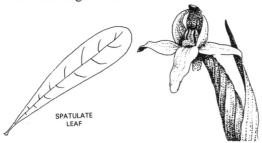

Figure 1173   **Figure 1174**

Spermatophyte. Plants reproducing by seeds.

Spermophyte. See **spermatophyte.**

Sphenoid. Wedge-shaped; cuneate. Figure 1175.

Spherical. A three-dimensional, isodiametrical structure, round in outline. Figure 1176. (same as **globose**)

Spheroidal. Almost spherical, but elliptical in cross

Figure 1175

section. Figure 1177.

Spicate. Arranged in a spike. Figure 1178; spike-like.

Spiciform. An inflorescence with the general appearance, but not necessarily the structure, of a true spike.

Spicula (pl. **spiculae**). See **spicule.**

Spicular. See **spiculate.**

Spiculate. Spicule-like; bearing spicules. Figure 1179.

Spicule. A short, pointed, epidermal projection. Figure 1179.

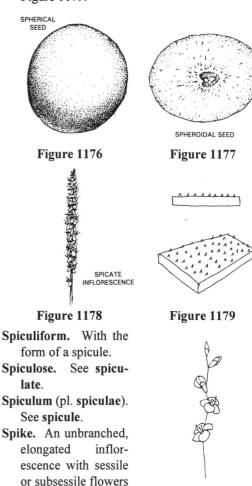

Figure 1176   **Figure 1177**

Figure 1178   **Figure 1179**

Spiculiform. With the form of a spicule.

Spiculose. See **spiculate.**

Spiculum (pl. **spiculae**). See **spicule.**

Spike. An unbranched, elongated inflorescence with sessile or subsessile flowers or spikelets maturing from the bottom upwards. Figure 1180. (compare **raceme**)

Figure 1180

Spikelet. A small spike or secondary spike; the ultimate flower cluster of grasses and sedges, consisting of 1-many flowers subtended by two bracts (glumes). Figure 1181.

Spindle-shaped. Broadest near the middle and tapering toward both ends, as in some roots. (see **fusiform**)

Spine. A stiff, slender, sharp-pointed structure arising from below the epidermis, representing a modified leaf or stipule; any structure with the appearance of a true spine. Figure 1182.

Spinescent. Bearing a spine or a spinelike point at the tip; bearing spines. Figure 1182.

Spiniferous. See **spinose**.

Spinose. Bearing spines. Figure 1183.

Spinous. See **spinose**.

Spinule. A small spine. Figure 1184.

Spinulose. Bearing spinules. Figure 1184.

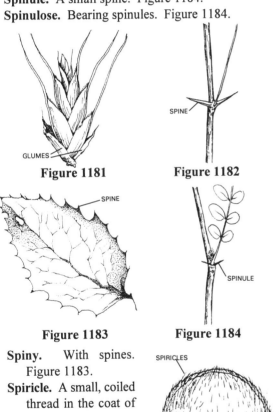

Figure 1181

Figure 1182

Figure 1183

Figure 1184

SPIRICLES

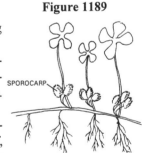

Figure 1185

Spiny. With spines. Figure 1183.

Spiricle. A small, coiled thread in the coat of some seeds and achenes which uncoils when moistened. Figure 1185.

Spongiose. Soft and spongy.

Sponsalia. See **an-**

thesis.

Sporadic. Occurring in a scattered distribution rather than in a continuous range.

Sporangiophore. A stalk bearing sporangia. Figure 1186.

Sporangium (pl. **sporangia**). A spore-bearing case or sac. Figures 1186, 1187 and 1188.

Spore. A reproductive cell resulting from meiotic cell division in a sporangium, representing the first cell of the gametophyte generation. Figure 1189.

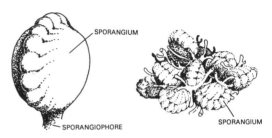

SPORANGIUM

SPORANGIOPHORE

SPORANGIUM

Figure 1186

Figure 1187

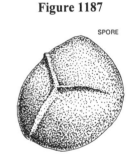

SPORE

SPOROPHYLL

SPORANGIA

Figure 1188

Figure 1189

Sporiferous. Bearing spores.

Sporocarp. A specialized structure containing sporangia. Figure 1190.

Sporophyll. A sporangium-bearing leaf, often modified in structure. Figure 1188.

SPOROCARP

Figure 1190

Sporophyte. The diploid (2n), spore-producing generation of the plant reproductive cycle, the dominant and conspicuous plant in the vascular plants. (compare **gametophyte**)

Sprawling. Bending or curving downward. Figure

1191; lying upon something and being supported by it.

Spray. A slender shoot or branch with its leaves, flowers, or fruits. Figure 1192.

Figure 1191   Figure 1192

Spreading. Extending nearly to the horizontal Figure 1193; almost prostrate.

Spumose. Frothy or foamy.

Spur. A hollow, slender, sac-like appendage of a petal or sepal, or of the calyx or corolla. Figure 1194; a short shoot bearing leaves or flowers and fruits. Figure 1195.

Spurred. Bearing a spur or spurs. Figures 1194 and 1195.

Squama (pl. **squamae**). A scale, as in some types of pappus in the Compositae (Asteraceae). Figure 1196.

Squamaceous. See **squamate**.

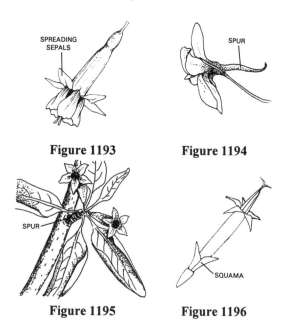

Figure 1193   Figure 1194

Figure 1195   Figure 1196

Squamate. Covered with scales (squamae). Figure 1197.

Squamella (pl. **squamellae**). A small scale or squama. Figure 1197.

Squamellate. With squamellae. Figure 1197.

Squamiform. Scale-like.

Squamose. See **squamate**.

Squamous. See **squamate**.

Squamule. The lodicule of a grass flower. Figure 1198.

Squamulose. With minute squamellae.

Squarrose. Abruptly recurved or spreading above the base; rough or scurfy due to the presence of recurved or spreading processes. Figure 1199.

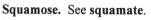
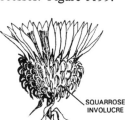

Figure 1197

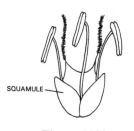

Figure 1198   Figure 1199

Squarrulose. With minute recurved processes.

Stalk. The supporting structure of an organ, usually narrower in diameter than the organ. Figure 1200.

Stamen (pl. **stamens, stamina**). The male reproductive organ of a flower, consisting of an anther and filament. Figure 1201; the angiosperm microsporophyll.

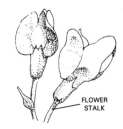

Figure 1200

Staminal. Of or pertaining to the stamens.

Staminate. Bearing stamens but not pistils, as a male flower which does not produce fruit or seeds. Figure 1202. (compare **pistillate**); bearing stamens.

Stamineal. See **staminal**.

Staminiferous. See **staminate**.

Staminode (pl. **staminodia**). A modified stamen which is sterile, producing no pollen. Figure 1203.

Staminodium. See **staminode**.

Staminody. A condition in which other organs, such as petals or sepals, become stamens.

Standard. The upper and usually largest petal of a papilionaceous flower, as in peas and sweet peas. Figure 1204.

Stelipilous. With stellate hairs. Figure 1206.

Stellate. Star-shaped, as in hairs with several to many branches radiating from the base. Figure 1206.

Stelliform. Star-shaped. Figure 1206.

Stem. The portion of the plant axis bearing nodes, leaves, and buds and usually found above ground. Figure 1207.

Stenopetalous. With narrow petals. Figure 1208.

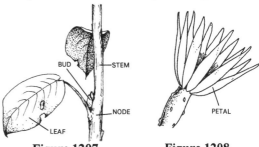

Figure 1207   **Figure 1208**

Stenophyllous. With narrow leaves. Figure 1209.

Steppe. Grassland; plain; prairie.

Stereomorphic. Radially symmetrical, so that a line drawn through the middle of the structure along any plane will produce a mirror image on either side; essentially the same as **actinomorphic**. Figure 1210.

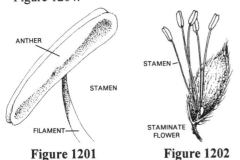

Figure 1201   **Figure 1202**

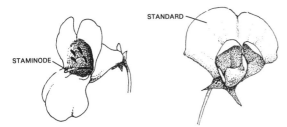

Figure 1203   **Figure 1204**

Stegium. A covering of thread-like hairs on the styles of some members of the Asclepiadaceae.

Stele. The primary vascular structure of a stem or root, including the vascular tissues and all tissues internal to the vascular tissues. Figure 1205.

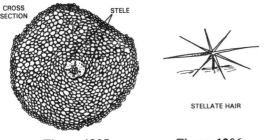

Figure 1205   **Figure 1206**

Figure 1209   **Figure 1210**

Sterigma. The persistent leaf base of the leaves of some coniferous trees. Figure 1211.

Sterile. Infertile, as a stamen that does not bear pollen, or a flower that does not bear seed. Figure 1212.

Figure 1211

Stigma. The portion of the pistil which is receptive to pollen. Figure 1213.

Stigmatic. Belonging to the stigma or having the characteristics of a stigma.

Stigmatiferous. Bearing a stigma. Figure 1213.

Stipe. A stalk supporting a structure, as the stalk attaching the ovary to the receptacle in some flowers. Figure 1214.

Stipel. A small, stipule-like structure at the base of a leaflet. Figure 1215.

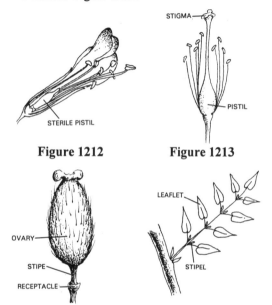

Figure 1212　　　　　**Figure 1213**

Figure 1214　　　　　**Figure 1215**

Stipellate. Bearing stipels. Figure 1215.

Stipellule. See **stipel**.

Stipitate. Borne on a stipe or stalk. Figure 1214.

Stipitate-glandular. Bearing stalked glands. Figure 1216.

Stipitiform. With the form of a stipe.

Stipular. Of or pertaining to a stipule.

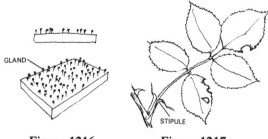

Figure 1216　　　　　**Figure 1217**

Stipulate. Bearing stipules. Figure 1217.

Stipule. One of a pair of leaf-like appendages found at the base of the petiole in some leaves. Figure 1217.

Stipuliform. Stipule-shaped.

Stipulose. See **stipulate**.

Stolon. An elongate, horizontal stem creeping along the ground and rooting at the nodes or at the tip and giving rise to a new plant. Figure 1218.

Stoloniferous. Bearing stolons. Figure 1218.

Stoloniform. Stolon-like.

Stoma. See **stomate**.

Stomate (pl. **stomata**). A pore or aperture, surrounded by two guard cells, which allows gaseous exchange. Figure 1219.

Stomatiferous. Bearing stomata. Figure 1219.

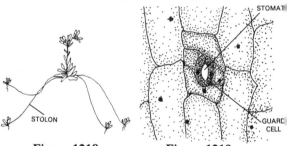

Figure 1218　　　　　**Figure 1219**

Stone. The hard, woody endocarp enclosing the seed of a drupe. Figure 1220.

Stone fruit. A drupe; a fruit with a stony pit. Figure 1220.

Stool. The base of plants which produce new stems each year. Figure 1221; a group of stems arising from a single root.

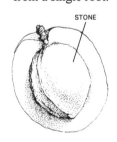

Figure 1220　　　　　**Figure 1221**

Stramineous. Straw-like in color or texture.

Strap. The ligule of a ray flower in the Compositae (Asteraceae). Figure 1222.

Strap-shaped. Elongated and flat. Figure 1222.

Streptocarpous. With twisted fruits. Figure 1223.

Stria (pl. **striae**). A fine line or groove. Figure 1224.

Striate. Marked with fine, usually parallel lines or grooves. Figure 1224.

Strict. Very straight and upright, not at all spreading. Figure 1225.

Striga. A bristle; a straight, stiff, sharp, appressed hair. Figure 1226.

Strigillose. Minutely strigose. Figure 1227.

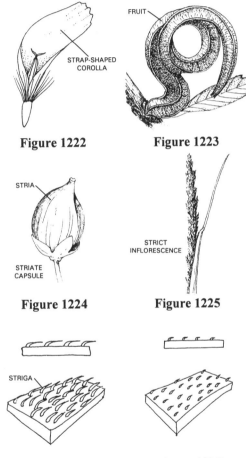

Figure 1222   **Figure 1223**

Figure 1224   **Figure 1225**

sporophylls on an axis. Figure 1229; a cone. Figure 1230.

Strombus. A legume which is spirally coiled, as in *Medicago*. Figure 1231.

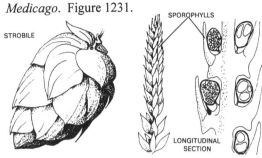

Figure 1228   **Figure 1229**

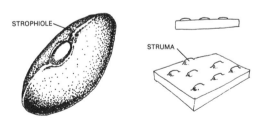

Figure 1230   **Figure 1231**

Strophiole. An appendage at the hilum in some seeds. Figure 1232.

Struma (pl. **strumae**). A cushion-like swelling. Figure 1233.

Strumose. With a covering of cushion-like swellings; bullate. Figure 1233.

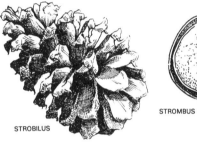

Figure 1226   **Figure 1227**

Figure 1232   **Figure 1233**

Strigose. Bearing straight, stiff, sharp, appressed hairs. Figure 1226.

Strigulose. See **strigillose**.

Strobilaceous. Of or pertaining to a cone; cone-like.

Strobile. A cone or an inflorescence resembling a cone, as in hops. Figure 1228.

Strobilus (pl. **strobili**). A cone-like cluster of

Stylar. Of or pertaining to a style.

Style. The usually narrowed portion of the pistil connecting the stigma to the ovary. Figure 1234.

Stylocarpellous. With a style, but without a stipe. Figure 1235. (compare **astylocarpellous**, and see **stylocarpepodic**)

Stylocarpepodic. With a style and a stipe. Figure

1236. (compare **astylocarpepodic**, and see **stylocarpellous**)

Stylodious. See **unicarpellous**.

Stylodium (pl. **stylodia**). A stigma branch, as in some members of the Geraniaceae. Figure 1237.

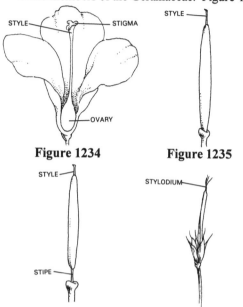

Figure 1234   **Figure 1235**

Figure 1236   **Figure 1237**

Stylopod. See **stylopodium**.

Stylopodic. With a stylopodium. Figure 1238.

Stylopodium. A disklike expansion or enlargement at the base of the style in the Umbelliferae (Apiaceae). Figure 1238.

Suaveolent. Fragrant.

Sub- (prefix). Meaning under, slightly, somewhat, or almost.

Subacute. Slightly acute. Figure 1239.

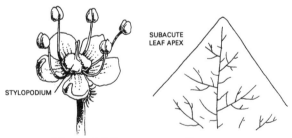

Figure 1238   **Figure 1239**

Subalpine. Growing in the mountains below the alpine zone and above the montane zone.

Subapical. Near the apex. Figure 1240.

Sub-basal. Near the base. Figure 1241.

Subcapitate. Almost capitate. Figure 1242.

Subcordate. Almost cordate. Figure 1243.

Subcorymbose. Almost corymbose.

Subcylindric. Almost cylindric in shape. Figure 1244.

Subentire. Almost entire. Figure 1245.

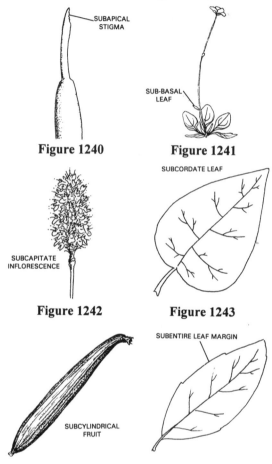

Figure 1240   **Figure 1241**

Figure 1242   **Figure 1243**

Figure 1244   **Figure 1245**

Suber. Cork.

Suberose. Corky in texture.

Suberous. See **suberose**.

Subfoliaceous. Almost foliaceous.

Subglabrate. Almost glabrous. Figure 1246.

Sublignous. Almost woody.

Submersed. Submerged.

Suborbicular. Almost orbicular in shape. Figure 1247.

Subrhizomatous. Almost rhizomatous.

SUBORBICULAR LEAF

Figure 1246   **Figure 1247**

Subscapose. Almost scapose. Figure 1241.

Subshrub. A suffrutescent perennial plant; a small shrub.

Subspicate. Almost spicate, but with short pedicels on some or all of the flowers or florets. Figure 1248.

SUBSPICATE INFLORESCENCE

Figure 1248

Subtend. To be below and close to, as a bract may subtend an inflorescence. Figure 1249.

Subterete. Almost terete. Figure 1250.

SUBTENDING BRACT

Figure 1249

SUBTERETE POD

Figure 1250

Subterranean. Below the surface of the ground.

Subterraneous. See **subterranean**.

Subtropical. Distributed in areas intermediate between tropical and temperate regions; nearly tropical.

Subula. A fine, sharp point. Figure 1251.

Subulate. Awl-shaped. Figure 1251.

SUBULATE LEAF

Figure 1251

Succulent. Juicy and fleshy, as the stem of a cactus or the leaves of *Aloe*. Figures 1252 and 1253.

Sucker. A shoot originating from below ground. Figure 1254.

Suffrutescent. Somewhat shrubby; slightly woody at the base.

Suffruticose. Somewhat woody.

Suffruticulose. See **suffruticose**.

Suffused. Tinted or tinged.

Sulcate. With longitudinal grooves or furrows. Figure 1255.

Sulcus (pl. **sulci**). A groove or furrow. Figure 1255.

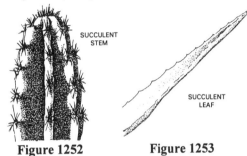

SUCCULENT STEM

SUCCULENT LEAF

Figure 1252   **Figure 1253**

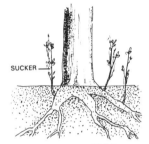

SUCKER

Figure 1254

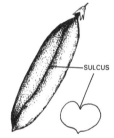

SULCUS

Figure 1255

Sulfureous. Sulfur-colored.

Summer annual. A plant with seeds germinating in spring or early summer and completing flowering and fruiting in late summer or early fall and then dying.

Superaxillary. Attached above the axil.

Superior. Attached above, as an ovary that is attached above the point of attachment of the other floral whorls. Figure 1256. (compare **inferior**)

ANDROECIUM   COROLLA

OVARY   CALYX

Figure 1256

Supra-axillary. See **superaxillary**.

Supraligular. Attached above the ligule.

Surculose. Producing suckers or runners from the base or from rootstocks. Figure 1257.

Surculose-proliferous. Reproducing by suckers or runners. Figure 1257.

Surculum. A fern rhizome. Figure 1258.

Surcurrent. Extending upward from the point of insertion, as a leaf base that extends up along the stem. Figure 1259.

Surficial. Growing near the ground, or spread over the surface of the ground. Figure 1260.

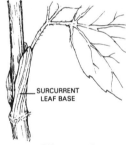

Figure 1257

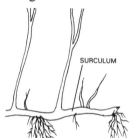

Figure 1258

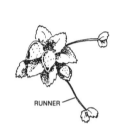

Figure 1259

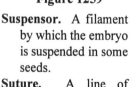

Figure 1260

Suspensor. A filament by which the embryo is suspended in some seeds.

Suture. A line of fusion; the line of dehiscence of a fruit or anther. Figure 1261.

Syconium (pl. **syconia**). The fruit of a fig, consisting of an entire ripened inflorescence with a hollow, inverted receptacle bearing flowers internally. Figure 1262.

Symmetric. Said of a flower having the same num-

Figure 1261

ber of parts in each floral whorl. Figure 1263.

Sympatric. Occupying the same geographic region. (compare **allopatric**)

Sympetalous. With the petals united, at least near the base. Figure 1264. (same as **gamopetalous**; compare **apopetalous** and **polypetalous**)

Symphysis. The fusion or coalescence of like parts as in a sympetalous corolla. Figure 1264.

Sympodial. Of or pertaining to a sympodium; in the form of a sympodium.

Sympodium. A main axis appearing to be simple, but actually consisting of a number of short axillary branches rather than a continuation of the main axis. Figure 1265. (compare **monopodium**)

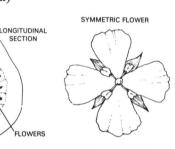

Figure 1262

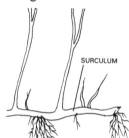

Figure 1263

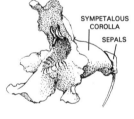

Figure 1264

Figure 1265

Syn-, Sym- (prefix). Meaning united.

Synandrous. With united anthers. Figure 1266.

Synantherous. See **synandrous**.

Synanthesis. The male and female parts of a flower maturing simultaneously. (compare **protandry**

Figure 1266

and **protogyny**)

Synanthious. With leaves appearing simultaneously with the flowers.

Synanthous. See **synanthious**.

Synanthy. The abnormal fusion of two or more flowers.

Syncarp. A multiple fruit. Figure 1267; an aggregate fruit. Figure 1268.

Syncarpous. Of or pertaining to a syncarp; with united carpels. Figure 1269. (compare **apocarpous**)

Synema. The column composed of united filaments in a flower with monadelphous stamens. Figure 1270.

Figure 1267   **Figure 1268**

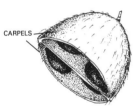

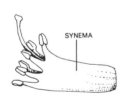

Figure 1269   **Figure 1270**

Syngenesious. With stamens united by their anthers. Figure 1266.

Synobasic. With a united base.

Synoecious. With staminate and pistillate flowers together in the same head. Figure 1271.

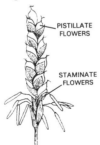

Figure 1271

Synsepalous. With united sepals. Figure 1264. (same as **gamosepalous**; compare **polysepalous**)

Tailed. With a tail-like appendage or appendages. Figure 1272.

Taproot. The main root axis from which smaller root branches arise; a root system with a main root axis and smaller branches, as in most dicots. Figure 1273. (compare **fibrous roots**)

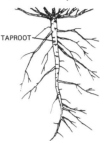

Figure 1272   **Figure 1273**

Tassel. The staminate inflorescence in corn (*Zea*). Figure 1274.

Tawny. Tan in color.

Taxon (pl. **taxa**). A taxonomic entity of any rank, such as order, family, genus, or species.

Tectum. The outermost layer of a pollen grain. Figure 1275.

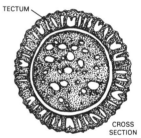

Figure 1274   **Figure 1275**

Tegule. One of the bracts of the involucre in the Compositae (Asteraceae). Figure 1276.

Temperate. Distributed in those regions of the earth lying between the tropic of Cancer (23 ½ degrees north latitude) and the Arctic Circle

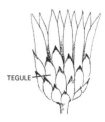

Figure 1276

(66 ⅔ degrees north latitude) or between the tropic of Capricorn (23 ½ degrees south latitude)

and the Antarctic Circle (66 ⅔ degrees south latitude).

Tendril. A slender, twining organ used to grasp support for climbing. Figure 1277.

Tendril-pinnate. Pinnately compound, but ending in a tendril, as in the sweet pea. Figure 1277.

Tentacle. A sensitive filament, as the glandular hairs of *Drosera*. Figure 1278.

Tentacular. Bearing tentacles. Figure 1278.

Tenuous. Slender or thin.

Tepal. A segment of a perianth which is not differentiated into calyx and corolla; a sepal or petal. Figure 1279.

Terete. Round in cross section; cylindrical. Figure 1280.

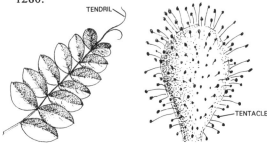

| **Figure 1277** | **Figure 1278** |

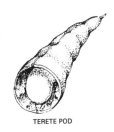

Figure 1279   **Figure 1280**

Tergeminate. Thrice divided into equal pairs; paired leaflets ternately compound. Figure 1281.

Terminal. At the tip or apex.

Ternary. Consisting of threes or involving threes; triple.

Ternate. In threes, as a leaf which is divided into three leaflets. Figure

Figure 1281

1282.

Terrestrial. Growing on ground; not aquatic.

Tesselate. With a checkered pattern. Figure 1283.

Testa (pl. **testae**). The seed coat, from the integuments of the ovule. Figure 1284.

Testaceous. Brick-red or brownish-red in color.

Testiculate. Resembling testicles. Figure 1285.

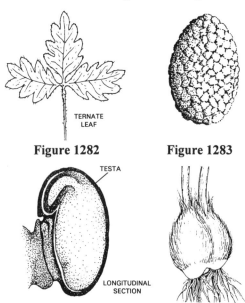

Figure 1282   **Figure 1283**

Figure 1284   **Figure 1285**

Tetra- (prefix). Meaning four.

Tetracyclic. With four whorls. Figure 1286.

Tetrad. A group of four.

Tetradinous. Occurring in tetrads.

Tetradymous. With four cells.

Tetradynamous. Having four long and two short stamens, as in most members of the Cruciferae (Brassicaceae). Figures 1287 and 1288.

Tetragonal. Four-angled. Figure 1289.

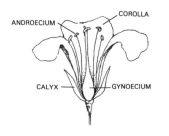

Figure 1286   **Figure 1287**

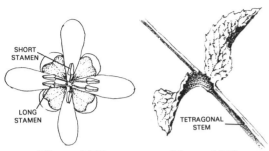

Figure 1288   **Figure 1289**

Tetrahedral. Four-sided, each side triangular. Figure 1290.

Tetramerous. With parts arranged in sets or multiples of four. Figure 1291.

Tetrandrous. With four stamens. Figure 1291.

Tetrangular. With four angles. Figure 1289.

Tetrapetalous. With four petals. Figure 1291.

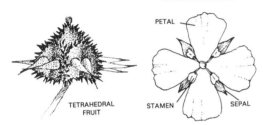

Figure 1290   **Figure 1291**

Tetraploid. With four representatives of each type of chromosome, or four complete sets of chromosomes, in each cell; 4x. (compare **diploid** and **haploid**)

Tetrapterous. With four wings or wing-like appendages. Figure 1292.

Tetrasepalous. With four sepals. Figure 1291.

Tetrastachyous. With four spikes.

Tetrastichous. In four vertical ranks or rows on an axis. Figure 1293.

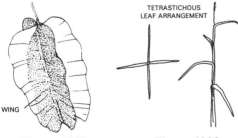

Figure 1292   **Figure 1293**

Thalamous. See **thalamus**.

Thalamus. The receptacle of a flower. Figure 1294.

Thalloid. Consisting of a thallus; resembling a thallus.

Thallus (pl. **thalli**). A plant body which is not obviously differentiated into stems, roots, and leaves. Figure 1295.

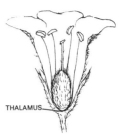

Figure 1294

Theca (pl. **thecae**). A pollen sac or cell of the anther. Figure 1296.

Thecate. With a theca. Figure 1296.

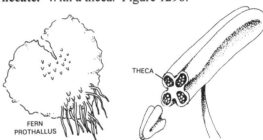

Figure 1295   **Figure 1296**

Thelephoroid. See **thelephorous**.

Thelephorous. With nipple-like protuberances. Figure 1297.

Thorn. A stiff, woody, modified stem with a sharp point; sometimes applied to any structure resembling a true thorn. Figure 1298. (compare **spine** and **prickle**)

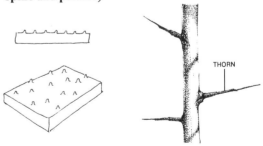

Figure 1297   **Figure 1298**

Three-ranked. In three vertical ranks or rows around an axis. Figure 1299.

Throat. The orifice of a gamopetalous corolla or gamosepalous calyx. Figure 1300; the expanded portion of the corolla between the limb and the tube. Figure 1301; the upper margin of the leaf sheath in grasses. Figure 1302.

Thrum. A heterostylic flower with a fairly short style and long stamens. Figure 1303. (compare **pin**)

Thyrse. A compact, cylindrical, or ovate panicle with an indeterminate main axis and cymose sub-axes. Figure 1304.

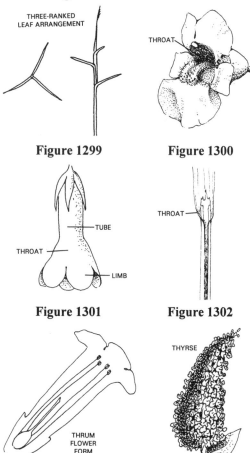

Figure 1299   **Figure 1300**

Figure 1301   **Figure 1302**

Figure 1303   **Figure 1304**

Thyrsoid. Thyrse-like.

Thyrsula. A small cyme borne in the leaf axil, as in many members of the Labiatae (Lamiaceae). Figure 1305.

Thyrsus. See **thyrse**.

Tiller. A basal or subterranean shoot which is more or less erect. Figure 1306. (compare **stolon** and **rhizome**)

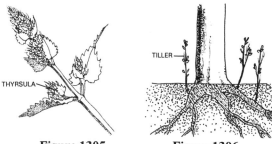

Figure 1305   **Figure 1306**

Tillering. A type of vegetative reproduction accomplished by tiller production. Figure 1306.

Tissue. A group of cells organized to perform a specific function, as epidermal tissue or vascular tissue. Figure 1307.

Tolerant. Capable of growing in the shade.

Tomentellous. See **tomentulose**.

Tomentose. With a covering of short, matted or tangled, soft, wooly hairs; with tomentum. Figure 1308. (compare **lanate** and **canescent**)

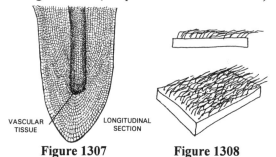

Figure 1307   **Figure 1308**

Tomentulose. Slightly tomentose. Figure 1309.

Tomentum (pl. **tomenta**). A covering of short, soft, matted, wooly hairs. Figure 1308.

Tongue. Ligule. Figure 1310.

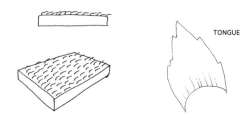

Figure 1309   **Figure 1310**

Tooth. Any small lobe or point along a margin. Figure 1311.

Toothed. Dentate. Figure 1311.

Torose. Cylindrical with alternate swellings and contractions. Figure 1312.

Tortuous. Twisted or bent. Figure 1313.

Torulose. Slightly torose, as in a small fruit which is constricted between the seeds. Figure 1314.

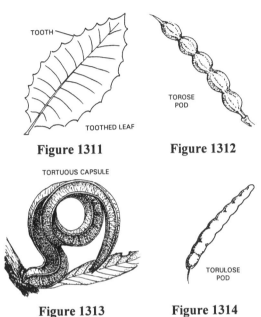

Figure 1311

Figure 1312

Figure 1313

Figure 1314

Torus (pl. **tori**). The receptacle of a flower. Figure 1315.

Trabecula (pl. **trabeculae**). A structure resembling a beam or crossbar. Figure 1316.

Trabecular. Of or pertaining to trabeculae.

Trabeculate. With a crossbar. Figure 1316.

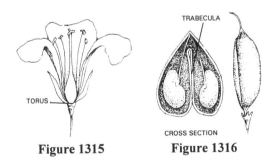

Figure 1315

Figure 1316

Trace. A vein. Figure 1317.

Tracheid. A xylem cell which is long, slender, and tapered at the ends. Figure 1318.

Trachycarpous. Rough-fruited.

Trachyspermous. Rough-seeded.

Trailing. Prostrate and creeping but not rooting.

Figure 1319.

Transcorrugated. Corrugated transversely to the axis. Figure 1320.

Translator. The connecting structure between the pollinia of adjacent anthers in the Asclepiadaceae. Figure 1321.

Translucent. Almost transparent. Figure 1322.

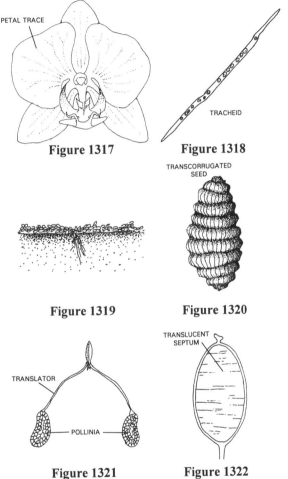

Figure 1317

Figure 1318

Figure 1319

Figure 1320

Figure 1321

Figure 1322

Transpiration. Emission of water vapor from the leaves, primarily through the stomata.

Transverse. At a right angle to the longitudinal axis of a structure. Figures 1323 and 1324.

Tree. A large woody plant, usually with a single main stem or trunk. Figure 1325.

Tri- (prefix). Meaning three.

Triachaenium. A fruit consisting of three achenes.

Triad. A group of three. Figure 1326.

Triadelphous. With stamens arranged into three

groups. Figure 1326.

Triandrous. With three stamens. Figure 1327.

Triangulate. Three-angled. Figure 1328.

Triaristate. Three-awned. Figure 1329.

Tricamarous. With three locules. Figure 1330.

Tricarinate. With three ridges or keels. Figure 1331.

Tricarpellary. With three carpels. Figure 1332.

Trichasium. A cymose inflorescence with three branches. Figure 1333.

Trichocarpous. With hairy fruit. Figure 1334.

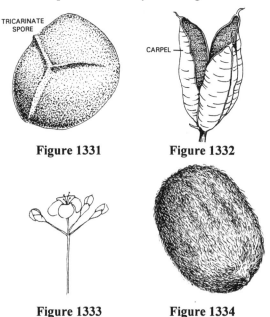

Figure 1331   **Figure 1332**

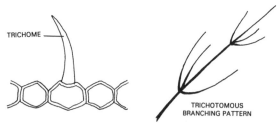

Figure 1333   **Figure 1334**

Trichome. A hair or hair-like outgrowth of the epidermis. Figure 1335.

Trichotomous. Three-forked. Figure 1336.

Tricolor. With three colors.

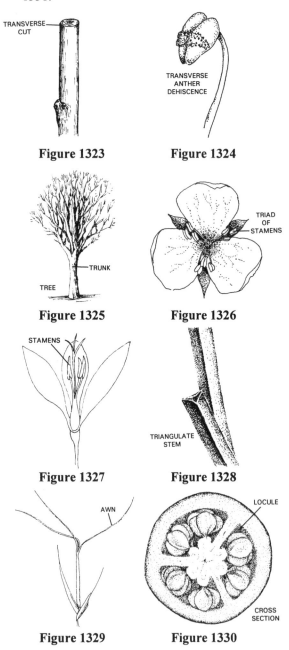

Figure 1323   **Figure 1324**

Figure 1325   **Figure 1326**

Figure 1327   **Figure 1328**

Figure 1329   **Figure 1330**

Figure 1335   **Figure 1336**

Tricussate. With whorls of three leaves, each alternating with leaves at the nodes above and below. Figure 1337.

Tricyclic. With three whorls, as in a flower with calyx, androecium, and gynoecium. Figure 1338.

Tridentate. Three-toothed. Figure 1339.

Tridigitate. Divided into three finger-like lobes or divisions. Figure 1340.

Tridynamous. With stamens arranged in two

groups of three, one group often longer than the other. Figure 1341.

Triecious. See **trioecious**.

Trifid. Three-cleft. Figure 1342.

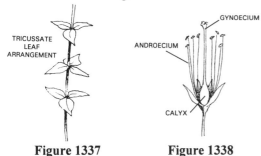

Figure 1337   **Figure 1338**

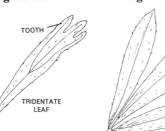

Figure 1339   **Figure 1340**

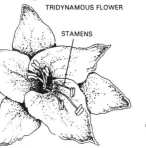

Figure 1341   **Figure 1342**

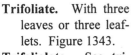

Figure 1343

Trifoliate. With three leaves or three leaflets. Figure 1343.

Trifoliolate. See **trifoliate**.

Trifurcate. Three-forked; divided into three branches. Figure 1344.

Trigamous. With three kinds of flowers, as in a plant with staminate, pistillate, and perfect flowers.

Trigeminate. With three pairs of leaflets. Figure 1345.

Trigeminous. See **trigeminate**.

Trigonal. See **trigonous**.

Trigonous. Three-angled. Figure 1346.

Trijugate. See **trigeminate**.

Trilobate. With three lobes. Figure 1342.

Trilocular. With three locules. Figure 1347.

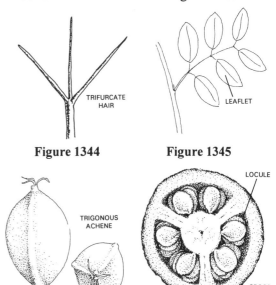

Figure 1344   **Figure 1345**

Figure 1346   **Figure 1347**

Trimerous. With parts arranged in sets or multiples of three. Figure 1348.

Trimonoecious. With male, female, and bisexual flowers on the same plant.

Trimorphic. With three forms.

Trinervate. See **trinerved**.

Trinerved. Three-nerved, with the nerves all arising from near the base. Figure 1349. (compare **triplinerved**)

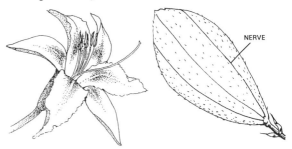

Figure 1348   **Figure 1349**

Trioecious. With male, female, and bisexual flowers on different plants.

Tripalmate. Palmately compound three times. Figure 1350.

Tripartite. Three-parted. Figure 1351.

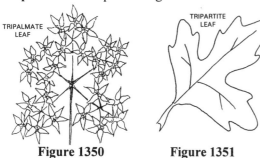

Figure 1350   **Figure 1351**

Tripetalous. With three petals. Figure 1352.

Triphyllous. With three leaves.

Tripinnate. Pinnately compound three times, with pinnate pinnules. Figure 1353.

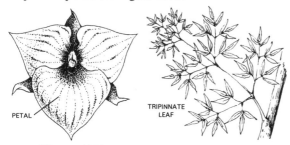

Figure 1352   **Figure 1353**

Tripinnatifid. Thrice pinnately cleft. Figure 1354.

Triple-nerved. See **triplinerved**.

Triplinerved. Three-nerved, with the two lateral nerves arising from the midnerve above the base. Figure 1355. (compare **trinerved**)

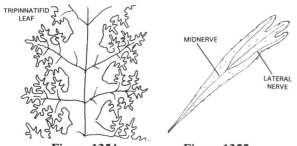

Figure 1354   **Figure 1355**

Tripterous. With three wings or wing-like appendages. Figure 1356.

Triquetrous. Three-edged; with three protruding

angles. Figure 1357.

Trispermous. Three-seeded.

Tristichous. In three vertical ranks or rows; three-ranked. Figure 1358.

Tristylous. With three styles. Figure 1359.

Trisulcate. With three furrows or grooves. Figure 1360.

Triternate. Triply ternate. Figure 1361.

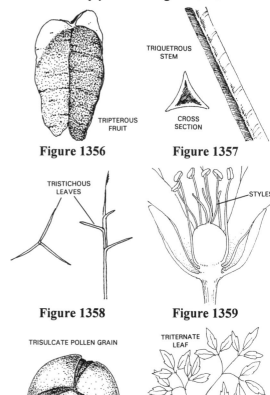

Figure 1356   **Figure 1357**

Figure 1358   **Figure 1359**

Figure 1360   **Figure 1361**

Tropical. Distributed in the tropics (i.e. between the tropic of Cancer and the tropic of Capricorn, or between 23 ½ degrees north latitude and 23 ½ degrees south latitude).

Truncate. With the apex or base squared at the end as if cut off. Figure 1362.

Trunk. The main stem of a tree below the branches. Figure 1363.

Tryma. A drupe-like nut with a fleshy, dehiscent

exocarp, as a walnut or hickory nut. Figure 1364.

Tube. A hollow, cylindrical structure, as the constricted basal portion of some gamopetalous corollas. Figure 1365.

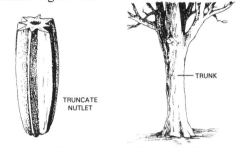

Figure 1362   **Figure 1363**

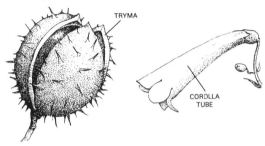

Figure 1364   **Figure 1365**

Tuber. The thickened portion of a rhizome bearing nodes and buds; underground stem modified for food storage. Figure 1366.

Tubercle. A small tuber-like swelling or projection. Figures 1367 and 1368; the base of the style in some members of the Cyperaceae. Figure 1369.

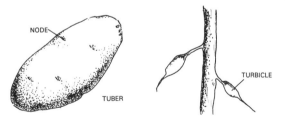

Figure 1366   **Figure 1367**

Tubercular. Of or pertaining to tubercles; tubercle-like.

Tuberculate. Bearing tubercles.

Tuberculation. See **tubercle**.

Tubercule. A nodule, as on the roots of some legumes. Figure 1370.

Figure 1368   **Figure 1369**

Tuberiferous. See **tuberculate**.

Tuberiform. Resembling a tuber.

Tuberoid. A thickened root which resembles a tuber.

Tuberous. Resembling a tuber; producing tubers.

Tubular. With the form of a tube or cylinder. Figure 1371.

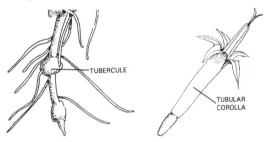

Figure 1370   **Figure 1371**

Tubuliflorous. Having tubular corollas in the perfect flowers of a head, as in some members of the Compositae (Asteraceae). Figure 1372.

Tubulous. With tubular flowers. Figure 1372.

Tufted. Arranged in a dense cluster. Figure 1373.

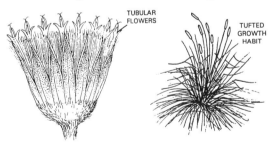

Figure 1372   **Figure 1373**

Tumescent. Somewhat tumid; swelling. Figure 1374.

Tumid. Swollen. Figure 1375.

Tunic. An integument; the outer coating of a seed or bulb. Figure 1376.

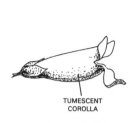

Figure 1374

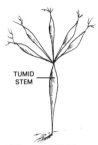

Figure 1375

Tunicate. Arranged in sheathing, concentric layers, as the leaves of an onion bulb. Figure 1377.

Turbinate. Top-shaped. Figure 1378.

Turgescent. See **turgid**.

Turgid. Swollen; expanded or inflated. Figure 1379.

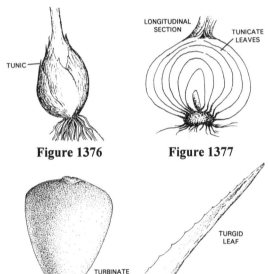

Figure 1376   **Figure 1377**

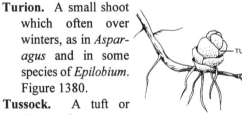

Figure 1378   **Figure 1379**

Turion. A small shoot which often over winters, as in *Asparagus* and in some species of *Epilobium*. Figure 1380.

Tussock. A tuft or clump of grasses or sedges. Figure 1373.

Twig. A small shoot or branch from a tree.

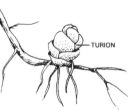

Figure 1380

Figure 1381.

Twining. Coiling or spiraling around a support (usually another stem) for climbing. Figure 1382.

Figure 1381   **Figure 1382**

Two-ranked. In two vertical ranks or rows on opposite sides of an axis; distichous. Figure 1383.

Type. The specimen that serves as the basis for a plant name.

Ubiquitous. Widespread; occurring in a wide variety of habitats.

Ultimate. The final section or division of a structure. Figure 1384.

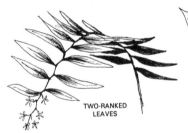

Figure 1383   **Figure 1384**

Umbel. A flat-topped or convex inflorescence with the pedicels arising more or less from a common point, like the struts of an umbrella. Figures 1385 and 1386; a highly condensed raceme.

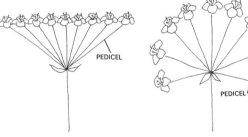

Figure 1385   **Figure 1386**

Umbellate. In umbels; umbel-like.

Umbellet. An ultimate umbellate cluster of a compound umbel. Figure 1387.

Umbelliferous. Bearing umbels; pertaining or belonging to the Umbelliferae (Apiaceae) family.

Figure 1387

Umbelliform. An inflorescence with the general appearance, but not necessarily the structure, of a true umbel. The term is often applied to inflorescences which are condensed cymes rather than condensed racemes.

Umbellule. See **umbellet**.

Umbilicate. With a depression in the middle, like a navel. Figure 1388.

Umbilicus. A navel-like structure, as the hilum of a seed. Figure 1389.

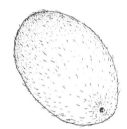

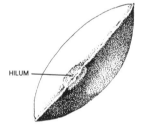

Figure 1388   **Figure 1389**

Umbo. A blunt or rounded protuberance, as on the ends of the scales of some pine cones. Figure 1390.

Umbonate. Possessing an umbo. Figure 1390.

Umbonulate. With a very small umbo. Figure 1391.

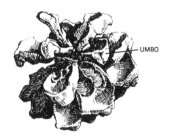

Figure 1390   **Figure 1391**

Umbraculate. Umbrella-shaped. Figure 1392.

Umbraculiferous. Bearing umbrella-shaped structures. Figure 1392.

Umbraculiform. See **umbraculate**.

Unarmed. Lacking spines, prickles, or thorns.

Uncate. See **uncinate**.

Uncinate. Hooked at the tip. Figure 1393.

Figure 1392   **Figure 1393**

Unctuous. Greasy; oily.

Undate. See **undulate**.

Undershrub. See **subshrub**.

Undulate. Wavy, but not so deeply or as pronounced as sinuate. Figure 1394. (Same as **repand**.)

Unequally pinnate. See **odd-pinnate**.

Unguicular. See **unguiculate**.

Unguiculate. Clawed. Figure 1395.

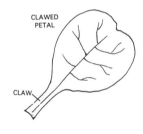

Figure 1394   **Figure 1395**

Ungulate. See **unguiculate**.

Uni- (prefix). Meaning one.

Uniaperturate. With a single aperture or opening. Figure 1396.

Uniaristate. One-awned. Figure 1397.

Uniaxial. With a single

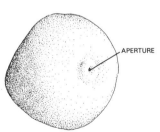

Figure 1396

unbranched stem terminating in a flower. Figure 1398.

Unicarpellate. See **unicarpellous**.

Unicarpellous. With a single, free carpel. Figure 1399.

Unicostate. With a single obvious rib, as in some leaves. Figure 1400.

Uniflorous. With a single flower. Figure 1398.

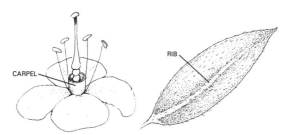

Figure 1397   **Figure 1398**

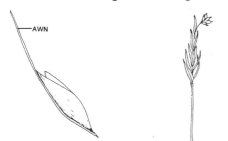

Figure 1399   **Figure 1400**

Unifoliate. With a single leaf; unifoliolate. Figure 1401.

Unifoliolate. A leaf theoretically compound, though only expressing a single leaflet and appearing simple, as in *Cercis*. Figure 1402.

Figure 1401   **Figure 1402**

Unijugate. A leaf pinnately compound, but consisting of only two leaflets. Figure 1403.

Unilateral. One-sided, as in an inflorescence with the flowers all on one side of the axis. Figure 1404.

Unilocular. With a single locule or compartment, as in some ovaries. Figure 1405.

Uniovulate. With a single ovule. Figure 1406.

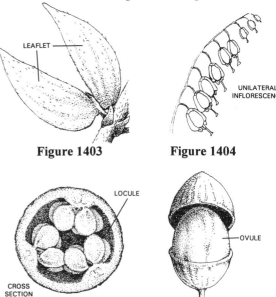

Figure 1403   **Figure 1404**

Figure 1405   **Figure 1406**

Uniparous. With only a single axis produced at each branching, as in some cymes. Figure 1407.

Unipetalous. With only a single petal.

Uniseptate. With only one septum, as in a silicle or silique. Figure 1408.

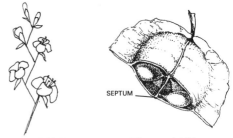

Figure 1407   **Figure 1408**

Uniseriate. Arranged in a single row or series. Figure 1409.

Unisexual. A flower with either male or female reproductive parts, but not both. The term is also applied to plants possessing such flowers. Figure 1410. (compare **bisexual** and **perfect**)

Urceolate. Pitcher-like; hollow and contracted near the mouth like a pitcher or urn. Figure 1411.

Urceolus. The perigynium in *Carex*. Figure 1412; any structure resembling a small pitcher.

Urceus. Any structure resembling a pitcher. Figures 1411 and 1412.

Urent. Stinging. Figure 1413.

Urn. The basal portion of a pyxis. Figure 1414.

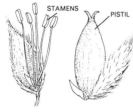

Figure 1409

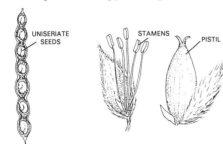

Figure 1410

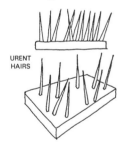

Figure 1411

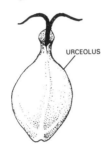

Figure 1412

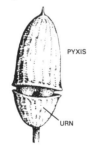

Figure 1413

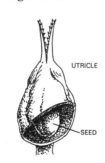

Figure 1414

Urn-shaped. See **urceolate**.

Utricle. A small, thin-walled, one-seeded, more or less bladdery-inflated fruit. Figure 1415.

Utricular. Of or pertaining to a utricle; bladder-like. Figure 1415.

Vagina. A sheath, as

Figure 1415

the sheathing petiole in grasses. Figure 1416.

Vaginate. Sheathed. Figure 1416.

Vaginiferous. Bearing sheaths.

Vallecula (pl. **valleculae**). A furrow, groove or depression. Figure 1417.

Vallecular. Of or pertaining to the valleculae.

Valleculate. Having valleculae. Figure 1417.

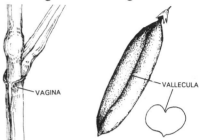

Figure 1416   **Figure 1417**

Valvate. Opening by valves, as in many dehiscent fruits. Figure 1418; a flower with the petals or sepals edge to edge along their entire length, but not overlapping. Figure 1419.

Valve. One of the segments of a dehiscent fruit, separating from other such segments at maturity. Figure 1418.

Figure 1418

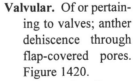

Figure 1419

Valvular. Of or pertaining to valves; anther dehiscence through flap-covered pores. Figure 1420.

Varicose. Swollen or enlarged in places. Figure 1421.

Variegated. Marked with patches or spots of different colors. Figure 1422.

Variety. A category in the taxonomic hierarchy

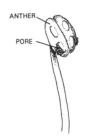

Figure 1420

below the species and subspecies level.

Vascular. Of or pertaining to conductive tissue (i.e., xylem and phloem); of or pertaining to plants possessing conductive tissue.

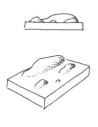 

Figure 1421   **Figure 1422**

VARIEGATED LEAF

Vascular bundle. A cluster or group of vascular tissues. Figure 1423.

Vegetative. Of or pertaining to the non-floral parts of a plant.

Vein. A vascular bundle, usually visible externally, as in leaves. Figure 1424.

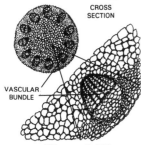 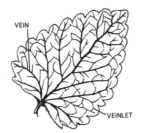

CROSS SECTION

VEIN

VASCULAR BUNDLE

VEINLET

Figure 1423   **Figure 1424**

Veinlet. A small vein. Figures 1424 and 1425.

Velamen (pl. **velamina**). The thick, spongy integument layer on the roots of some epiphytic orchids.

Velum. The membranous flap of tissue partially covering the sporangium of *Isoetes*. Figure 1426.

VEINLET

Figure 1425

Velumen. A covering of short, soft hairs. Figure 1427.

Velutinous. Velvety; covered with short, soft, spreading hairs. Figure 1427.

Venation. The pattern of veining on a leaf. Figure 1424.

Venenose. See **venomous**.

Venomous. Poisonous.

Venose. Veiny; venous. Figures 1424 and 1425.

Venous. Of or pertaining to veins; vein-like.

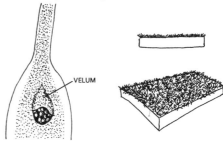

VELUM

Figure 1426   **Figure 1427**

Ventral. Pertaining to the front or inward surface of an organ in relation to the axis, as in the upper surface of a leaf; adaxial. Figure 1428. (compare **dorsal**)

Ventricose. Inflated or swollen on one side only, as in some corollas, especially in the genus *Penstemon*. Figure 1429.

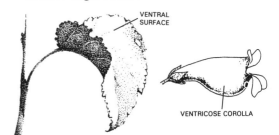

VENTRAL SURFACE

VENTRICOSE COROLLA

Figure 1428   **Figure 1429**

Venulation. Diminutive of venation, referring to the pattern of veinlets on a leaf. Figure 1424.

Venule. See **veinlet**.

Venulose. With veinlets. Figures 1424 and 1425.

Venulous. See **venulose**.

Vermicular. See **vermiform**.

Vermiform. Worm-shaped. Figure 1430.

Vermiformous. See **vermiform**.

Vernal. Flowering or appearing in the spring.

Vernation. The arrangement of leaves within the bud.

Vernicose. Shiny; with a varnished appearance and texture.

Verrucose. Warty; covered with wart-like elevations. Figure 1431.

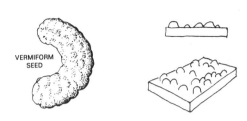

Figure 1430   **Figure 1431**

Verruculose. Covered with very small wart-like elevations. Figure 1432.

Versatile. Attached near the middle rather that at one end, as some anthers. Figure 1433. (compare **basifixed** and **dorsifixed**)

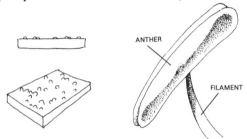

Figure 1432   **Figure 1433**

Versicolor. Of various colors; changeable in color.

Vertical. Positioned lengthwise, in the same direction as the axis; leaves positioned with the blade perpendicular, so that neither surface is obviously the upper or lower. Figure 1434.

Verticil. An arrangement of similar parts around a central axis or point of attachment; a whorl. Figure 1435.

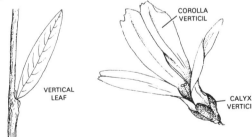

Figure 1434   **Figure 1435**

Verticillaster. A pair of axillary cymes arising from opposite leaves or bracts and forming a false whorl. Figure 1436.

Verticillate. Arranged in verticils; whorled. Figure

1437.

Vesicle. A small bladder-like structure. Figure 1438.

Vesicular. Of or pertaining to vesicles.

Vespertine. Opening, or functioning, in the evening.

Vessel. A tube-like xylem structure composed of vessel elements attached end to end.

Vessel element. A short, thick xylem cell with blunt ends. Figure 1439.

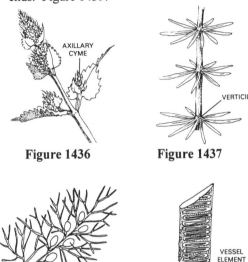

Figure 1436   **Figure 1437**

Figure 1438   **Figure 1439**

Vessel member. See **vessel element**.

Vestigial. An organ or structure which is much reduced and likely nonfunctional, though believed at one time to have been more perfectly formed. Figures 1440, 1441, and 1442. (see **rudimentary** and **obsolete**)

Figure 1440   **Figure 1441**

Vestiture. The epidermal coverings of a plant, collectively.

Vesture. See **vestiture**.

Vexillum. The upper and usually largest petal of a papilionaceous flower, as in peas and sweet peas. Figure 1443.

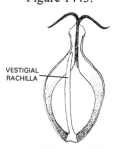
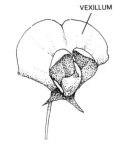

Figure 1442　　　　**Figure 1443**

Villose. See **villous**.

Villosulous. Diminutive if **villous**.

Villous. Bearing long, soft, shaggy, but unmatted, hairs. Figure 1444.

Villus (pl. **villi**). A long, soft, shaggy hair. Figure 1444.

Vimineous. With long, flexible twigs; composed of twigs; twig-like.

Vinaceous. Wine-colored.

Vine. A plant with the stem not self-supporting, but climbing or trailing on some support. Figure 1445.

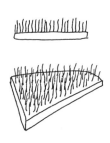

Figure 1444　　　　**Figure 1445**

Vinicolor. See **vinaceous**.

Violaceous. Violet-colored; of or pertaining to the Violaceae.

Virescence. The condition of becoming green.

Virescent. Becoming green; greenish.

Virgate. Wand-like; straight, slender, and erect. Figure 1446.

Viridescent. See **virescent**.

Viscid. Sticky or gummy.

Viscidium. A sticky structure on the pollinia of the Orchidaceae which attaches to a pollinator. Figure 1447.

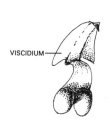

Figure 1446　　　　**Figure 1447**

Viscidulous. Slightly sticky.

Vitreous. Transparent.

Vitta (pl. **vittae**). An oil tube in the carpel walls of the fruits of the Umbelliferae (Apiaceae). Figure 1448.

Vittate. Having vittae. Figure 1448.

Figure 1448

Viviparous. Sprouting on the parent plant, as the bulblets forming in some inflorescences. Figure 1449.

Volute. Rolled up. Figure 1450.

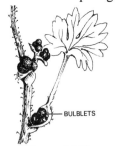
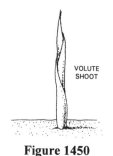

Figure 1449　　　　**Figure 1450**

Wart. A firm protuberance. Figure 1451.

Webbed. With an interlacing network of filaments, fibers, hairs, or veins. Figure 1452.

Weed. An aggressive plant which colonizes disturbed habitats and cultivated lands.

Whorl. A ring-like arrangement of similar parts arising from a common point or node; a verticil. Figure 1453.

Whorled. With parts arranged in whorls, as in a leaf

arrangement with three or more leaves arising from a node. Figure 1454. (Same as **verticillate**.)

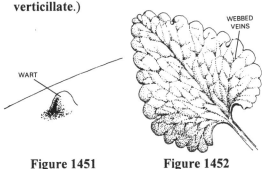

Figure 1451   **Figure 1452**

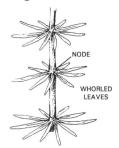

Figure 1453   **Figure 1454**

Wing. A thin, flat margin bordering or extending from a structure. Figures 1455 and 1456; one of the two lateral petals of a papilionaceous corolla. Figure 1457.

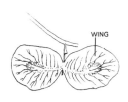

Figure 1455   **Figure 1456**

Winged. Possessing wings. Figures 1455, 1456, and 1457.

Winter annual. A plant with seeds germinating in late summer or fall and completing flowering and fruiting in spring or summer of the following

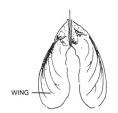

Figure 1457

year and then dying.

Winter bud. A hibernating vegetative shoot.

Woolly. With long, soft, entangled hairs; lanate. Figure 1458.

X. When placed before a specific epithet, indicates the taxon is of known hybrid origin.

Xanthic. Yellowish.

Xanthophyll. Yellow, orange, or red fat-soluble pigments.

Xenogamy. Pollination between flowers of separate plants.

Xeric. Of dry areas.

Xero- (prefix). Meaning dry.

Xeromorphic. Possessing obvious physical adaptations for a dry environment, such as the succulent, water storing stem of a cactus.

Xerophilous. See **xeric**.

Xerophyte (adj. **xerophytic**). A plant adapted to life in dry environments.

Xylem. The water conducting tissue of vascular plants. Figure 1459.

Zonate. Marked or colored in circular rings or zones. Figure 1460.

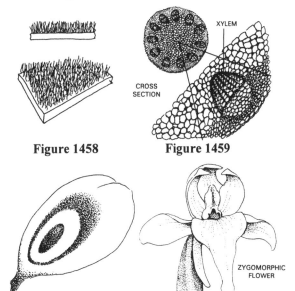

Figure 1458   **Figure 1459**

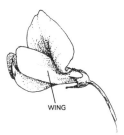

Figure 1460   **Figure 1461**

Zoophilous. Animal-pollinated.

Zygomorphic. Bilaterally symmetrical, so that a line drawn through the middle of the structure along only one plane will produce a mirror image on either side. Figure 1461. (compare **actino-morphic**, and see **irregular**)

Zygomorphous. See **zygomorphic**.

PART TWO

TERMINOLOGY BY CATEGORY

ROOTS

That portion of the plant axis lacking nodes and leaves and usually found below ground.

ROOT PARTS

Cortex. Root tissue between the epidermis and the stele. Figure 1462.

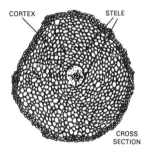

Figure 1462

Meristem. Undifferentiated, actively dividing tissues at the growing tips of shoots and roots. Figure 1463.

Pith. The spongy, parenchymatous central tissue in some roots. Figure 1464.

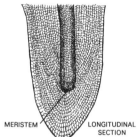

Figure 1463

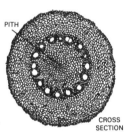

Figure 1464

Rootlet. A small root. Figure 1465.

Stele. The primary vascular structure of a root. Figure 1462.

Tubercule. A nodule, as on the roots of some legumes. Figure 1466.

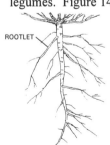

Figure 1465

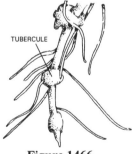

Figure 1466

ROOT SHAPES

Fusiform. Spindle-shaped; broadest near the middle and tapering toward both ends. Figure 1467.

Napiform. Turnip-shaped. Figure 1468.

Rapiformis. See **napiform**.

Spindle-shaped. See **fusiform**.

Turbinate. Top-shaped. Figure 1469.

Figure 1467

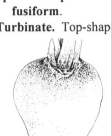

Figure 1468   **Figure 1469**

ROOT TYPES

Adventitious. Structures or organs developing in an unusual position, as roots originating on the stem. Figure 1470.

Aerial. Occurring above ground or water.

Buttressed. With props or supports, as in the flared trunks of some trees. Figure 1471.

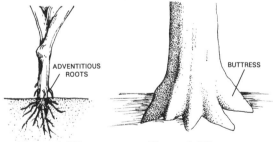

Figure 1470   **Figure 1471**

Fibrous roots. A root system with all of the

branches of approximately equal thickness, as in the grasses and other monocots. Figure 1472.

Haustorium (pl. **haustoria**). A specialized root-like organ used by parasitic plants to draw nourishment from host plants.

Figure 1472

Prop root. Adventitious roots arising from lower nodes and providing support to a stem. Figure 1473.

Radicant. A root arising from the node of a prostrate stem. Figure 1474.

Figure 1473

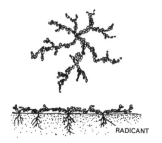

Figure 1474

Rhizoid. A root-like structure lacking conductive tissues (xylem and phloem). Figure 1475.

Rootlet. A small root. Figure 1476.

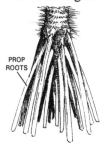

Figure 1475

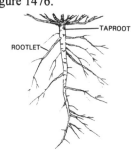

Figure 1476

Surculum. A fern rhizome. Figure 1477.

Taproot. The main root axis from which smaller root branches arise; a root system with a main root axis and smaller branches, as in most dicots. Figure

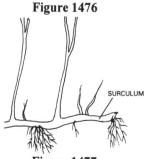

Figure 1477

1476.

Tuberoid. A thickened root which resembles a tuber. Figure 1478.

Tubercule. A nodule, as on the roots of some legumes. Figure 1479.

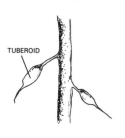

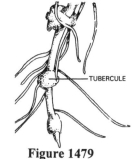

Figure 1478   **Figure 1479**

STEMS

The portion of the plant axis bearing nodes, leaves, and buds and usually found above ground.

STEM PARTS

Bark. The outermost layers of a woody stem including all of the living and nonliving tissues external to the cambium. Figure 1480.

Bole. See **trunk**.

Branchlet. A small branch. Figure 1481.

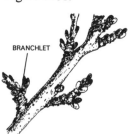

Figure 1480   **Figure 1481**

Bud. An undeveloped shoot or flower. Figure 1482.

Bundle scar. Scar left on a twig by the vascular bundles when a leaf falls. Figure 1482.

Cambium. A tissue composed of cells capable of active cell division, producing xylem to the inside of the plant and phloem to the outside; a lateral meristem. Figure 1483.

Caudex (pl. **caudices, caudexes**). The persistent

and often woody base of a herbaceous perennial. Figure 1484.

Cortex. Bark or rind; root tissue between the epidermis and the stele. Figure 1483.

Crown. The persistent base of a herbaceous perennial, a caudex. Figure 1484; the top part of a tree. Figure 1485.

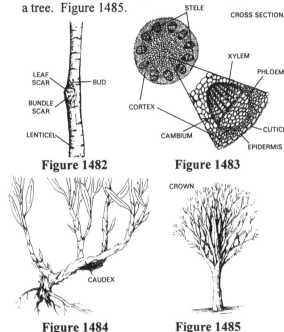

Figure 1482   **Figure 1483**

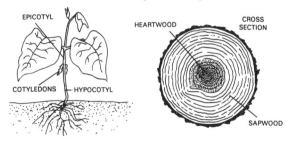

Figure 1484   **Figure 1485**

Cuticle. The waxy layer on the surface of a stem. Figure 1483.

Epicotyl. That portion of the embryonic stem above the cotyledons. Figure 1486.

Epidermis. The outermost cellular layer of a stem. Figure 1483.

Heartwood. The innermost, usually somewhat darker wood of a woody stem. Figure 1487.

Figure 1486   **Figure 1487**

Hypocotyl. That portion of the embryonic stem below the cotyledons. Figure 1486.

Internode. The portion of a stem between two nodes. Figure 1488.

Joint. The section of a stem from which a leaf or branch arises; a node, especially on a grass stem. Figure 1489.

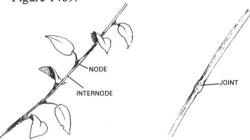

Figure 1488   **Figure 1489**

Leaf scar. The scar remaining on a twig after a leaf falls. Figure 1482.

Lenticel. A slightly raised, somewhat corky, often lens-shaped area on the surface of a young stem. Figure 1482.

Meristem. Undifferentiated, actively dividing tissues at the growing tip of a shoot.

Node. The position on the stem where leaves or branches originate. Figure 1490.

Phloem. The food conducting tissue of vascular plants; bark. Figure 1483.

Pith. The spongy, parenchymatous central tissue in some stems and roots. Figure 1491.

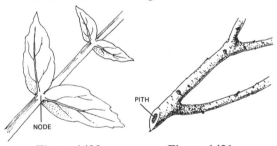

Figure 1490   **Figure 1491**

Prickle. A small, sharp outgrowth of the epidermis or bark. Figure 1492.

Sapwood. The outer, newer, usually somewhat lighter, wood of a woody stem; the wood that is actively transporting water; alburnum. Figure 1487.

Stele. The primary vascular structure of a stem, including the vascular tissues and all tissues internal to the vascular tissues. Figure 1483.

Trunk. The main stem of a tree below the branches. Figure 1493.

Twig. A small shoot or branch from a tree. Figure 1494.

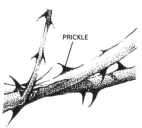

Figure 1492

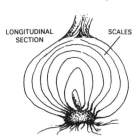

Figure 1493

Figure 1494

STEM TYPES

Bulb. An underground bud with thickened fleshy scales, as in the onion. Figure 1495.

Bulbel. A small bulb arising from the base of a larger bulb. Figure 1496.

Figure 1495

Figure 1496

Bulbil. See **bulblet**.

Bulblet. A small bulb.

Caudex (pl. **caudices**, **caudexes**). The persistent and often woody base of a herbaceous perennial. Figure 1497.

Caulicle. A small stem; a rudimentary stem.

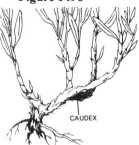

Figure 1497

Caulis. The main stem of a herbaceous plant. Figure 1498.

Cladode. See **cladophyll**.

Cladophyll. A stem with the form and function of a leaf. Figures 1499 and 1500.

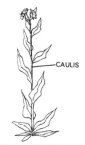

Figure 1498

Figure 1499

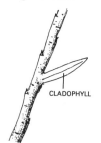

Figure 1500

Corm. A short, solid, vertical underground stem with thin papery leaves. Figure 1501.

Cormel. A small corm arising at the base of a larger corm.

Culm. A hollow or pithy stalk or stem, as in the grasses, sedges, and rushes. Figure 1502.

Figure 1501

Floricane. The second-year flowering and fruiting cane (shoot) of *Rubus*. (compare **primocane**)

Liana. A woody, climbing vine.

Monopodium (pl. **monopodia**). A single main axis giving rise to lateral branches. Figure 1503.

Figure 1502

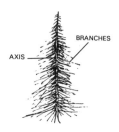

Figure 1503

Offset. A short, often prostrate, shoot originating near the ground at the base of another shoot. Figure 1504.

Offshoot. A shoot or branch arising from a main stem. Figure 1505.

Figure 1504   **Figure 1505**

Phylloclade. See **cladophyll**.

Primocane. The first-year, usually flowerless, cane (shoot) of *Rubus*. (compare **floricane**)

Pseudoscape. A false scape, where not all of the leaves are truly basal in origin though, superficially, they appear to be so. Figure 1506.

Ratoon. A shoot arising from the root of a plant that has been cut down. Figure 1507.

Figure 1506   **Figure 1507**

Rhizome. A horizontal underground stem; rootstock. Figure 1508.

Runner. A slender stolon or prostrate stem rooting at the nodes or at the tip. Figure 1509.

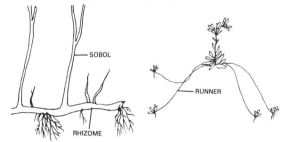

Figure 1508   **Figure 1509**

Sarment. A long, slender runner. See **runner**.

Shoot. A young stem or branch. Figure 1510.

Sobol. Elongated caudex branches; a shoot arising from the base of a stem or from the rhizome. Figure 1508.

Sobole. See **sobol**.

Spray. A slender shoot or branch with its leaves, flowers, or fruits. Figure 1511.

Figure 1510   **Figure 1511**

Stolon. An elongate, horizontal stem creeping along the ground and rooting at the nodes or at the tip and giving rise to a new plant. See **runner**.

Stool. The base of plants which produce new stems each year. See **caudex**; a group of stems arising from a single root.

Sucker. A shoot originating from below ground. Figure 1512.

Sympodium. A main axis appearing to be simple, but actually consisting of a number of short axillary branches rather than a continuation of the main axis. Figure 1513.

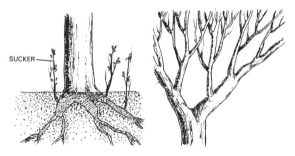

Figure 1512   **Figure 1513**

Thorn. A stiff, woody, modified stem with a sharp point; sometimes applied to any structure resembling a true thorn. Figure 1514.

Tiller. A basal or subterranean shoot which is more or less erect. See **sucker**.

Trunk. The main stem of a tree below the branches.

Figure 1515.

Tuber. The thickened portion of a rhizome bearing nodes and buds; underground stem modified for food storage. Figure 1516.

Turion. A small shoot which often over winters, as in some species of *Epilobium*. Figure 1517.

Figure 1514   **Figure 1515**

Figure 1516   **Figure 1517**

Twig. A small shoot or branch from a tree. Figure 1518.

Vine. A plant with the stem not self-supporting, but climbing or trailing on some support.

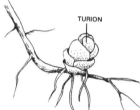

Figure 1518

STEM FORMS

Acaulescent. Without a stem, or the stem so short that the leaves are apparently all basal, as in the dandelion. Note: the peduncle should not be confused with the stem. Figure 1519.

Adscendent. See **ascending**.

Adsurgent. See **ascending**.

Alate. Winged. Figure 1520.

Ancipital. Two-edged, as the winged stem of *Sisyrinchium*. Figure 1520.

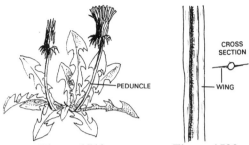

Figure 1519   **Figure 1520**

Angulate. Angled. Figure 1521.

Aphyllopodic. Having the lowermost leaves reduced to small scales. Figure 1522.

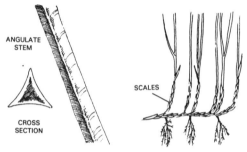

Figure 1521   **Figure 1522**

Aphyllous. Without leaves.

Ascendent. See **ascending**.

Ascending. Growing obliquely upward, usually curved. Figure 1523.

Assurgent. See **ascending**.

Caespitose. Growing in dense tufts. Figure 1524.

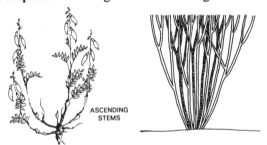

Figure 1523   **Figure 1524**

Caulescent. With an obvious leafy stem rising above the ground.

Cauliflorous. Bearing flowers on the stem or trunk. Figure 1525.

Cauline. Of, on, or pertaining to the stem.

Cespitose. See **caespitose**.

Clambering. Weakly climbing on other plants or

surrounding objects.

Climbing. Growing more or less erect by leaning or twining on another structure for support. Figure 1526.

Figure 1525 **Figure 1526**

Creeping. Growing along the surface of the ground, or just beneath the surface, and producing roots, usually at the nodes. Figure 1527.

Decumbent. Reclining on the ground but with the tip ascending. Figure 1528.

Figure 1527 **Figure 1528**

Deliquescent. An irregular pattern of branching without a well defined central axis from bottom to top. Figure 1529.

Dichotomous. Branched or forked into two more or less equal divisions. Figure 1530.

Figure 1529 **Figure 1530**

Divaricate. Widely diverging or spreading apart. Figure 1531.

Divergent. Diverging or spreading. Figure 1532.

Figure 1531 **Figure 1532**

Eramous. With unbranched stems. Figure 1533.

Erect. Vertical, not declining or spreading. Figure 1534.

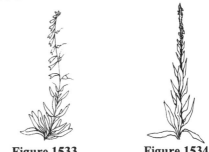

Figure 1533 **Figure 1534**

Fasciated. Compressed into a bundle or band; grown closely together; with the stems malformed and flattened as if several separate stems had been fused together. Figure 1535.

Fasciculate. Arranged in fascicles. Figure 1536.

Figure 1535 **Figure 1536**

Fastigiate. Clustered, parallel, and erect, giving a broom-like appearance. Figure 1524.

Fleshy. Thick and pulpy; succulent. Figure 1537.

Frutescent. Shrubby or shrub-like.

Fruticose. See **frutescent**.

Fruticulose. Somewhat shrubby; small and shrubby.

Herbaceous. Not woody.

Jointed. Having nodes or points of articulation, as in the stems of *Opuntia*. Figure 1538.

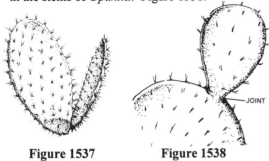

Figure 1537   **Figure 1538**

Macrocladous. With long branches.
Multicipital. With many heads, as in a highly branched caudex. Figure 1539.
Nodiferous. See **nodose**.
Nodose. With nodes. Figure 1540.

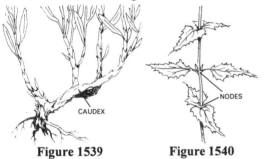

Figure 1539   **Figure 1540**

Nudicaul. With leafless stems.
Orthocladous. With straight branches. Figure 1541.
Orthotropic. Of, pertaining to, or exhibiting an essentially vertical growth habit. Figure 1542.

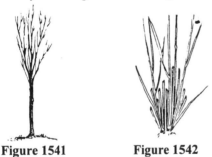

Figure 1541   **Figure 1542**

Pachycladous. With thick branches.
Pluricipital. See **multicipital**.
Procumbent. Lying or trailing on the ground, but not rooting at the nodes. Figure 1543.
Prostrate. Lying flat on the ground. Figure 1544.

Pterocaulous. With winged stems. Figure 1545.
Pulvinate. Cushion-like or mat-like. Figure 1546.

Figure 1543   **Figure 1544**

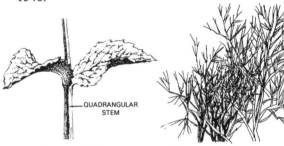

Figure 1545   **Figure 1546**

Pulviniform. See **pulvinate**.
Quadrangular. Four-angled. Figure 1547.
Quadrangulate. See **quadrangular**.
Ramiform. Branchlike in form; branched.
Ramose. With many branches; branching. Figure 1548.

Figure 1547   **Figure 1548**

Ramous. See **ramose**.
Ramulose. See **ramose**.
Reclining. Bending or curving downward. Figure 1549; lying upon something and being supported by it.
Recumbent. Leaning or resting on the ground; prostrate. Figure 1550.
Repent. Prostrate; creeping. Figure 1551.
Rhizomatous. With rhizomes. Figure 1552.

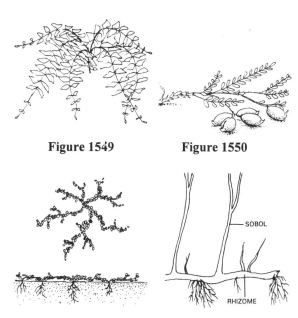

Figure 1549

Figure 1550

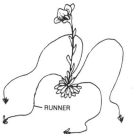

Figure 1556

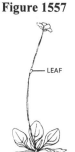

Figure 1557

Figure 1551

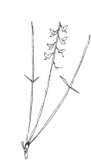

Figure 1552

Figure 1558.

Simple. Unbranched.

Soboliferous. Of or pertaining to sobols; bearing sobols. Figure 1552.

Sprawling. Bending or curving downward; lying upon something and being supported by it. See **reclining**.

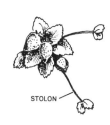

Figure 1558

Rosulate. With the leaves arranged in basal rosettes, the stem very short or lacking. Figure 1553.

Rush-like. Grass-like in appearance, with inconspicuous flowers. Figure 1554.

Spreading. Extending nearly to the horizontal; almost prostrate. Figure 1559.

Stoloniferous. Bearing stolons. Figure 1560.

Figure 1559

Figure 1560

Figure 1553

Figure 1554

Sarcocaulous. With fleshy stems. Figure 1555.

Sarcous. Fleshy. See **sarcocaulous**.

Sarmentose. With long, slender runners. Figure 1556.

Scandent. Climbing. Figure 1557.

Scapiform. Scape-like but not entirely leafless.

Figure 1555

Stoloniform. Stolon-like.

Strict. Very straight and upright, not at all spreading. Figure 1561.

Subscapose. Almost scapose. Figure 1558.

Subterranean. Below the surface of the ground.

Figure 1561

Subterraneous. See **subterranean**.

Succulent. Juicy and fleshy, as the stem of a cactus.

Figure 1562.

Suffrutescent. Somewhat shrubby; slightly woody at the base.

Suffruticose. Somewhat woody.

Suffruticulose. See **suffruticose**.

Surculose. Producing suckers or runners from the base or from rootstocks. Figure 1563.

Figure 1562

Surficial. Growing near the ground, or spread over the surface of the ground. Figure 1564.

Figure 1567   **Figure 1568**

Vimineous. With long, flexible twigs; composed of twigs; twig-like.

Virgate. Wand-like; straight, slender, and erect. Figure 1569.

Winged. Possessing wings. Figure 1570.

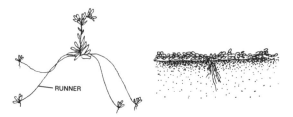

Figure 1563   **Figure 1564**

Tetragonal. Four-angled. Figure 1565.

Trailing. Prostrate and creeping but not rooting. Figure 1564.

Triangulate. Three-angled. Figure 1566.

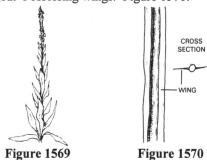

Figure 1569   **Figure 1570**

LEAVES

The usually expanded, photosynthetic organs of a plant.

LEAF PARTS

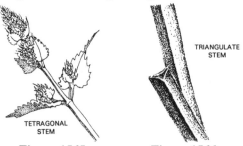

Figure 1565   **Figure 1566**

Trigonal. See **trigonous**.

Trigonous. Three-angled. Figure 1566.

Tufted. Arranged in a dense cluster.

Twining. Coiling or spiraling around a support (usually another stem) for climbing. Figure 1567.

Unarmed. Lacking spines, prickles, or thorns.

Uniaxial. With a single unbranched stem terminating in a flower. Figure 1568.

Apex. The tip; the point farthest from the point of attachment. Figure 1571.

Base. The end of the leaf blade nearest to the point of attachment. Figure 1571.

Blade. The broad part of a leaf. Figure 1571.

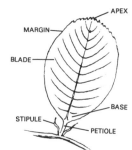

Figure 1571

Collar. The area on the outside of a grass leaf at the juncture of the blade and sheath. Figure 1572.

Costa (pl. **costae**). A rib or prominent midvein of a

leaf. See **midvein**.

Denticle. A small tooth or tooth-like projection. Figure 1573.

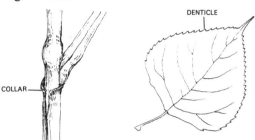

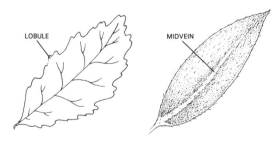

Figure 1572   **Figure 1573**

Leaflet. A division of a compound leaf. Figure 1574.

Limb. The expanded part of a leaf. See **blade**.

Lobe. A rounded division or segment of an organ, as of a leaf. Figure 1575.

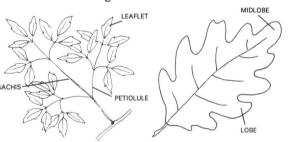

Figure 1574   **Figure 1575**

Lobule. A small lobe; a lobe-like subdivision of a lobe. Figure 1576.

Margin. The edge of a leaf blade. Figure 1571.

Midlobe. The central lobe of a leaf. Figure 1575.

Midnerve. The central nerve of a leaf. See **midvein**.

Midrib. The central rib or vein of a leaf. See **midvein**.

Midvein. The central vein of a leaf. Figure 1577.

Figure 1576   **Figure 1577**

Mucro. A short, sharp, abrupt point, usually at the tip of a leaf. Figure 1578.

Nerve. A prominent, simple vein or rib of a leaf.   See illustration for **midvein**.

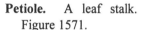

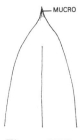

Figure 1578

Petiole. A leaf stalk. Figure 1571.

Petiolule. The stalk of a leaflet of a compound leaf. Figure 1574.

Pinna (pl. **pinnae**). One of the primary divisions or leaflets of a pinnate leaf. See illustration for **leaflet**.

Pinnule. The pinnate division of a pinna in a bipinnately compound leaf, or the ultimate divisions of a leaf which is more than twice pinnately compound. Figure 1579.

Rachis. The main axis of a compound leaf. Figure 1574.

Rhachis. See **rachis**.

Rib. A main longitudinal vein in a leaf. See **vein**.

Segment. A section or division of a leaf. Figure 1580.

Sinus. The cleft, depression, or recess between two lobes of a leaf. Figure 1580.

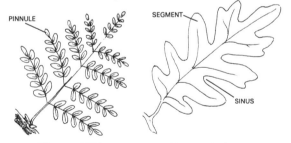

Figure 1579   **Figure 1580**

Stipule. One of a pair of leaf-like appendages found at the base of the petiole in some leaves. Figure 1571.

Tendril. A slender, twining organ used to grasp support for climbing. Figure 1581.

Tooth. Any small lobe or point along a margin. See **denticle** illustration.

Vein. A vascular bundle, usually visible externally, as in leaves. Figure 1582.

Veinlet. A small vein. Figure 1582.

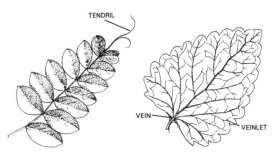

Figure 1581   Figure 1582

LEAF SHAPES (Figure 1583.)

Acerose. Needle-shaped, as the leaves of pine or spruce.

Awl-shaped. Short, narrowly triangular, and sharply pointed like an awl.

Cordate. Heart-shaped, with the notch at the base.

Deltoid. With the shape of the Greek letter delta; shaped like an equilateral triangle.

Elliptic. In the shape of an ellipse, or a narrow oval; broadest at the middle and narrower at the two equal ends.

Elliptical. See **elliptic**.

Ensiform. Sword-shaped, as an *Iris* leaf.

Falcate. Sickle-shaped; hooked; shaped like the beak of a falcon.

Flabellate. Fan-shaped.

Flabelliform. See **flabellate**.

Gladiate. Sword-shaped.

Halberd-shaped. See **hastate**.

Hastate. Arrowhead-shaped, but with the basal lobes turned outward rather than downward; halberd-shaped. (compare **sagittate**)

Lanceolate. Lance-shaped; much longer than wide, with the widest point below the middle.

Linear. Resembling a line; long and narrow with more or less parallel sides.

Lyrate. Lyre-shaped; pinnatifid, with the terminal lobe large and rounded and the lower lobes much smaller.

Obcordate. Inversely cordate, with the attachment at the narrower end; sometimes refers to any leaf with a deeply notched apex.

Obdeltoid. Deltoid, with the attachment at the pointed end.

Obelliptic or **obelliptical.** Almost elliptic, but with the distal end somewhat larger than the proximal end.

Oblanceolate. Inversely lanceolate, with the attachment at the narrower end.

Oblong. Two to four times longer than broad with nearly parallel sides.

Obovate. Inversely ovate, with the attachment at the narrower end.

Orbicular. Approximately circular in outline.

Orbiculate. See **orbicular**.

Oval. Broadly elliptic, the width over one-half the length.

Ovate. Egg-shaped in outline and attached at the broad end (applied to plane surfaces).

Pandurate. Fiddle-shaped.

Panduriform. See **pandurate**.

Peltate. Shield-shaped; borne on a stalk attached to the lower surface rather than to the base or margin.

Perfoliate. A leaf with the margins entirely surrounding the stem, so that the stem appears to pass through the leaf.

Quadrate. Square; rectangular.

Reniform. Kidney-shaped.

Rhombic. Diamond-shaped.

Rhomboid. See **rhomboidal**.

Rhomboidal. Quadrangular, nearly rhombic, with obtuse lateral angles.

Rotund. Round or rounded in outline.

Sagittate. Arrowhead-shaped, with the basal lobes directed downward. (compare **hastate**)

Spathulate. See **spatulate**.

Spatulate. Like a spatula in shape, with a rounded blade above gradually tapering to the base.

Subulate. Awl-shaped.

LEAF BASES (Figure 1584.)

Aequilateral. Equal-sided, as opposed to oblique.

Attenuate. Tapering gradually to a narrow base.

Auriculate. With ear-shaped appendages.

Cordate. Heart-shaped, with the notch at the base.

Cuneate. Wedge-shaped, triangular and tapering to a point at the base.

Eared. See **auriculate**.

Figure 1583a.   **Figure 1583b.**

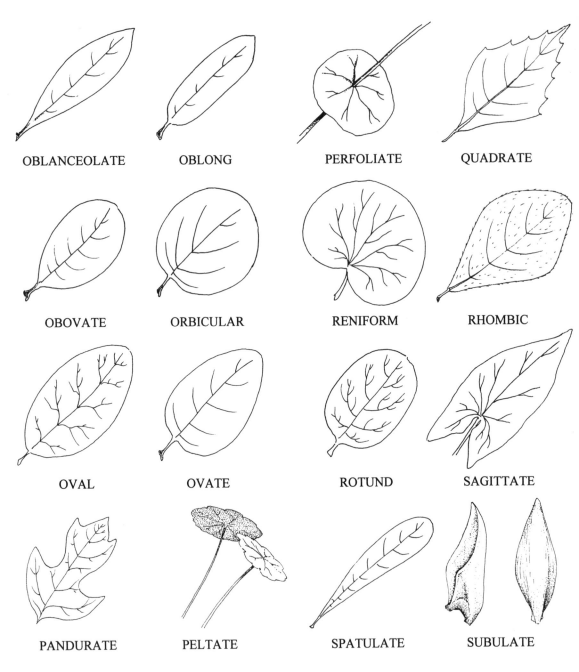

OBLANCEOLATE   OBLONG   PERFOLIATE   QUADRATE

OBOVATE   ORBICULAR   RENIFORM   RHOMBIC

OVAL   OVATE   ROTUND   SAGITTATE

PANDURATE   PELTATE   SPATULATE   SUBULATE

Figure 1583c.   **Figure 1583d.**

Halberd-shaped. See **hastate**.

Hastate. Arrowhead-shaped, but with the basal lobes turned outward rather than downward; halberd-shaped. (compare **sagittate**)

Inequilateral. See **oblique**.

Oblique. With unequal sides; slanting.

Rounded. With a rounded base.

Sagittate. Arrowhead-shaped, with the basal lobes directed downward. (compare **hastate**)

Truncate. With the base squared at the end as if cut off.

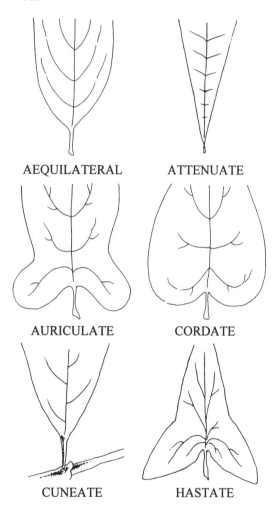

AEQUILATERAL   ATTENUATE

AURICULATE   CORDATE

CUNEATE   HASTATE

Figure 1584a.

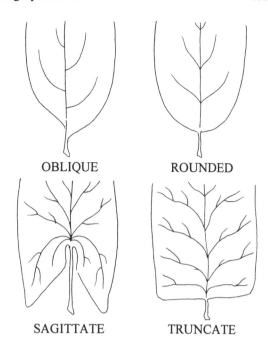

OBLIQUE   ROUNDED

SAGITTATE   TRUNCATE

Figure 1584b.

LEAF APICES   (Figure 1585.)

Abrupt. Terminating suddenly. See **truncate**.

Acuminate. Gradually tapering to a sharp point and forming concave sides along the tip.

Acute. Tapering to a pointed apex with more or less straight sides.

Apiculate. Ending abruptly in a small, slender point.

Aristate. Bearing an awn or bristle at the tip.

Aristulate. Bearing a minute awn or bristle at the tip.

Caudate. With a tail-like appendage.

Cirrhous. See **cirrose**.

Cirrose. With a cirrus (tendril).

Cuspidate. Tipped with a short, sharp, abrupt point (cusp).

Emarginate. With a notch at the apex.

Mucronate. Tipped with a short, sharp, abrupt point (mucro).

Mucronulate. Tipped with a very small mucro.

Muticous. Blunt, without a point or spine.

Obcordate. With a deeply notched apex.

Obtuse. Blunt or rounded at the apex; with the sides

coming together at the apex at an angle greater than 90 degrees.

Praemorse. Terminating abruptly, as if bitten off. See **truncate**.

Pungent. Tipped with a sharp, rigid point; with a sharp, acrid odor or taste.

Retuse. With a shallow notch in a round or blunt apex.

Rounded. With a rounded apex.

Subacute. Slightly acute.

Truncate. With the apex squared at the end as if cut off.

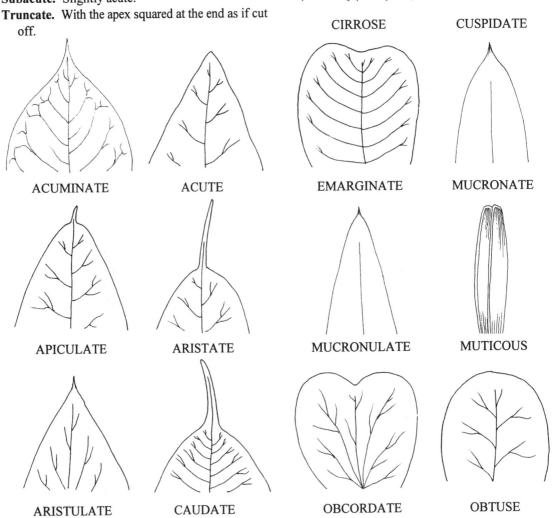

CIRROSE   CUSPIDATE

ACUMINATE   ACUTE   EMARGINATE   MUCRONATE

APICULATE   ARISTATE   MUCRONULATE   MUTICOUS

ARISTULATE   CAUDATE   OBCORDATE   OBTUSE

Figure 1585a.   **Figure 1585b.**

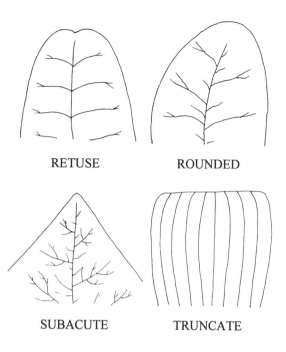

RETUSE   ROUNDED

SUBACUTE   TRUNCATE

Figure 1585c.

LEAF DIVISION   (Figure 1586.)

Abruptly pinnate. Pinnate without an odd leaflet at the tip. Same as **even-pinnate**.

Bifoliate. With two leaves or two leaflets.

Bipinnate. Twice pinnate; with the divisions again pinnately divided.

Biternate. Doubly ternate with the ternate divisions again ternately divided.

Compound leaf. A leaf separated into two or more distinct leaflets.

Decompound.   More than once-compound, the leaflets again divided.

Even-pinnate. Pinnately compound with a terminal pair of leaflets or a tendril rather than a single terminal leaflet, so that there is an even number of leaflets.

Foliolate. Pertaining to or having leaflets; usually used in compounds, such as **bifoliolate** or **trifoliolate**.

Imparipinnate. Odd-pinnate; unequally pinnate.

Interruptedly pinnate. Pinnate with leaflets of various sizes intermixed.

Key to Common Types of Leaf Divisions

1 Leaf blade not divided into separate leaflets. . . . . **Simple**
1 Leaf blade divided into separate leaflets. (**Compound**)
  2 Leaflets arising from a common point, like the fingers of a hand. (**Palmate**)
    3 Leaflets simple.
      4 Leaflets three. . . . . **Ternate, Trifoliate**
      4 Leaflets more than three. . . . . **Palmate**
    3 Leaflets divided into secondary leaflets. (**Decompound**)
      5 Leaves twice divided. . . . . **Biternate**
      5 Leaves thrice divided. . . . . **Triternate**
  2 Leaflets arising from opposite sides of an elongated axis. (**Pinnate**)
    6 Leaflets simple.
      7 Leaflets even in number.
        8 Leaf ending in a tendril. . . . . **Tendril-pinnate**
        8 Leaf not ending in a tendril. **Even-pinnate, Abruptly pinnate**
      7 Leaflets odd in number.
        9 Leaflets three. . . . . **Ternate, Trifoliate**
        9 Leaflets more than three. . . . . **Odd-pinnate, Unequally pinnate, Imparipinnate**
    6 Leaflets divided into secondary leaflets. (**Decompound**)
      10 Leaflets twice divided. . . . . **Bipinnate**
      10 Leaflets thrice divided. . . . . **Tripinnate**

Odd-pinnate. Pinnately compound with a terminal leaflet rather than a pair of leaflets or a tendril, so that there is an odd number of leaflets.

Palmate. Lobed, veined, or divided from a common point, like the fingers of a hand. (compare **pinnate**)

Pinnate. A compound leaf with leaflets arranged on opposite sides of an elongated axis. (compare **palmate**)

Simple. Undivided, as a leaf blade which is not separated into leaflets (though the blade may be deeply lobed or cleft).

Tendril-pinnate. Pinnately compound, but ending in a tendril, as in the sweet pea.

Ternate. In threes, as a leaf which is divided into three leaflets.

Trifoliate. With three leaves or three leaflets.

Trifoliolate. See **trifoliate**.

Tripinnate. Pinnately compound three times, with pinnate pinnules.

Triternate. Triply ternate.

Unequally pinnate. See **odd-pinnate**.

Unifoliate. A leaf theoretically compound, though only expressing a single leaflet and appearing simple, as in *Cercis*.

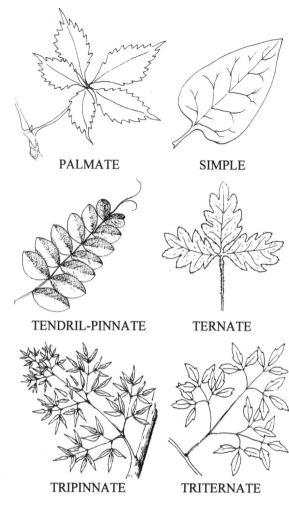

PALMATE   SIMPLE

TENDRIL-PINNATE   TERNATE

TRIPINNATE   TRITERNATE

Figure 1586b.

LEAF VENATION

The pattern of veining on a leaf.

Costate. Ribbed. Figure 1587.

Net-veined. In the form of a network; reticulate. Figure 1588.

Parallel-veined. With the main veins parallel to the leaf axis or to each other. Figure 1589. (compare **net-**

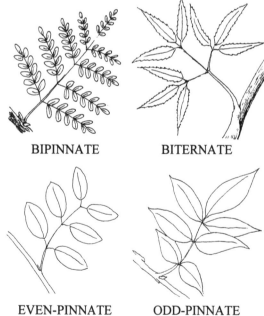

BIPINNATE   BITERNATE

EVEN-PINNATE   ODD-PINNATE

Figure 1586a.

Figure 1587

Figure 1588

Figure 1589

veined)

Pinnipalmate. Intermediate between pinnate and palmate, as in a leaf with the first pair of veins larger and more distinctive than the others. Figure 1590.

Figure 1590

Reticulate. In the form of a network; net-veined. Figure 1588.

Ribbed. With prominent ribs or veins. See illustration for **costate**.

Trinerved. Three-nerved, with the nerves all arising from near the base. Figure 1591. (compare **triplinerved**)

Triplinerved. Three-nerved, with the two lateral nerves arising from the midnerve above the base. Figure 1592. (compare **trinerved**)

Venose. Veiny. Figure 1588.

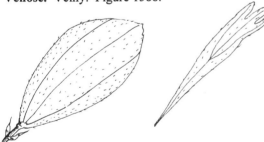

Figure 1591   **Figure 1592**

LEAF MARGINS (Figure 1593.)

Bidentate. With two teeth.

Bifid. Deeply two-cleft or two-lobed, usually from the tip.

Bilobed. Divided into two lobes.

Bipartite. Divided almost to the base into two divisions.

Bipinnatifid. Twice pinnately cleft.

Bisected. Split into two parts.

Biserrate. Doubly serrate, as when the teeth of a serrate leaf are also serrate.

Cleft. Cut or split about half-way to the middle or base.

Crenate. With rounded teeth along the margin.

Crenulate. With very small rounded teeth along the margin.

Crisped. Curled, wavy or crinkled.

Dentate. Toothed along the margin, the teeth directed outward rather than forward.

Denticulate. Dentate with very small teeth.

Digitate. Lobed, veined, or divided from a common point, like the fingers of a hand. (same as **palmate**)

Dissected. Deeply divided into many narrow segments.

Divided. Cut or lobed, essentially to the base or to the midrib.

Edentate. Without teeth.

Entire. Not toothed, notched, or divided, as the continuous margins of some leaves.

Erose. With the margin irregularly toothed, as if gnawed.

Erosulate. More or less erose.

Incised. Cut sharply, deeply, and usually irregularly.

Inrolled. Curled or rolled inward; involute.

Involute. With the margins rolled inward toward the upper side. (compare **revolute**)

Lacerate. Cut or cleft irregularly, as if torn.

Laciniate. Cut into narrow, irregular lobes or segments.

Lobed. Bearing lobes which are cut less than half way to the base or midvein.

Lobulate. With lobules.

Multifid. Cleft into many narrow segments or lobes.

Palmate. Lobed, veined, or divided from a common point, like the fingers of a hand. (compare **pinnate**)

Palmatifid. Palmately cleft or lobed.

Palmatisect. Palmately divided.

Parted. Deeply cleft, usually more than half the

distance to the base or midvein.

Pedate. Palmately divided, with the lateral lobes 2-cleft.

Pinnatifid. Pinnately cleft or lobed half the distance or more to the midrib, but not reaching the midrib.

Pinnatilobate. With pinnately arranged lobes.

Pinnatisect. Pinnately cleft to the midrib.

Quadripinnatifid. Four times pinnately cleft.

Repand. With a slightly wavy or weakly sinuate margin. Same as **undulate**.

Revolute. With the margins rolled backward toward the underside. (compare **involute**)

Runcinate. Sharply pinnatifid or cleft, the segments directed downward.

Serrate. Toothed along the margin, the sharp teeth pointing forward.

Serrulate. Toothed along the margin with minute, sharp, forward-pointing teeth.

Sinuate. With a strongly wavy margin.

Key to Common Leaf Margin Types

1 Margin continuous, not toothed, notched, lobed, or divided. . . . . **Entire**
1 Margin toothed, notched, lobed, or divided.
  2 Leaf toothed, notched, lobed, or incised less than half the distance to the base or midvein.
    3 Margin not distinctly toothed or lobed, merely wavy.
      4 Margin tightly wavy, producing a crinkled appearance. . . . . **Crisped**
      4 Margin loosely wavy, producing a smoother appearance. . . . . **Repand, Sinuate**
    3 Margin distinctly toothed or lobed.
      5 Leaf toothed or lobed only at the apex.
        6 Apex with two teeth or lobes. . . . . **Bidentate**
        6 Apex with three teeth or lobes. . . . . **Tridentate**
      5 Leaf toothed or lobed below the apex.
        7 Margin coarsely lobed or cleft.
          8 Margin irregularly cleft. . . . . **Cleft**
          8 Margin regularly lobed. . . . . **Lobed, Pinnatilobate**
        7 Margin finely toothed or lobed.
          9 Margin not sharply toothed. . . . . **Crenate**
          9 Margin sharply toothed.
            10 Margin irregularly toothed. . . . . **Erose**
            10 Margin regularly toothed.
              11 Teeth directed forward. . . . . **Serrate**
              11 Teeth directed outward. . . . . **Dentate**
  2 Leaf lobed, divided, or incised half or more the distance to the base or midvein.
    12 Leaf parted or divided only at the apex.
      13 Leaf divided at the apex into two sections. . . . . **Bipartite**
      13 Leaf divided at the apex into three sections. . . . . **Trifid**
    12 Leaf parted or divided below the apex.
      14 Leaf palmately lobed or divided.
        15 Leaf divisions resembling the fingers of a hand. . . . . **Digitate**
        15 Leaf divisions not particularly resembling the fingers of a hand. . . . . **Palmatifid**
      14 Leaf pinnately lobed or divided.
        16 Leaf irregularly cleft or divided.
          17 Leaf divided into many very narrow segments. . . . . **Dissected**
          17 Leaf divided more broadly.
            18 Leaf segments directed outward or forward. **Incised, Lacerate**
            18 Leaf segments directed backward. . . . . **Runcinate**
        16 Leaf regularly cleft or divided.
          19 Leaf cleft to the midvein. . . . . **Pinnatisect**
          19 Leaf not cleft to the midvein. . . . . **Pinnatifid**

Sinuous. Of a wavy or serpentine form. See illustration for **sinuate**.
Tridentate. Three-toothed.
Trifid. Three-cleft.

Tripartite. Three-parted.
Tripinnatifid. Thrice pinnately cleft.
Undulate. Wavy, but not so deeply or as pronounced as sinuate. See illustration for **repand**.

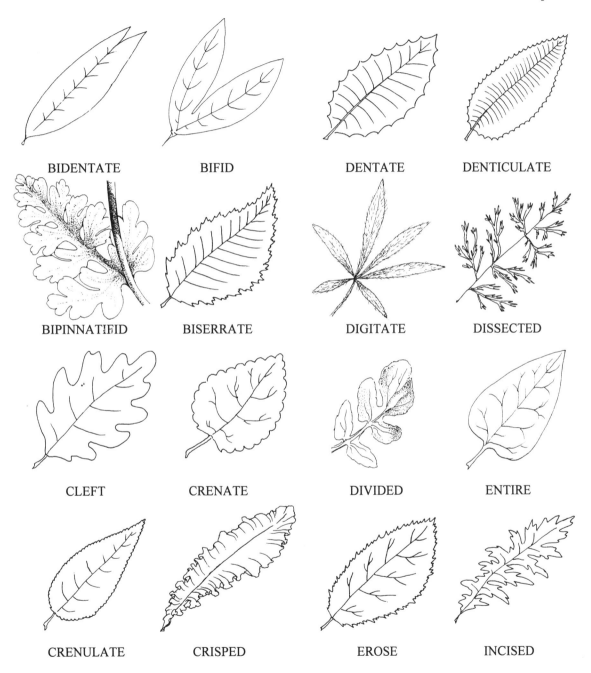

BIDENTATE   BIFID   DENTATE   DENTICULATE

BIPINNATIFID   BISERRATE   DIGITATE   DISSECTED

CLEFT   CRENATE   DIVIDED   ENTIRE

CRENULATE   CRISPED   EROSE   INCISED

Figure 1593a.   **Figure 1593b.**

INVOLUTE   LACERATE   PEDATE   PINNATIFID

LACINIATE   LOBED   PINNATILOBATE   PINNATISECT

LOBULATE   PALMATIFID   REPAND   REVOLUTE

PALMATISECT   PARTED   RUNCINATE   SERRATE

Figure 1593c.   **Figure 1593d.**

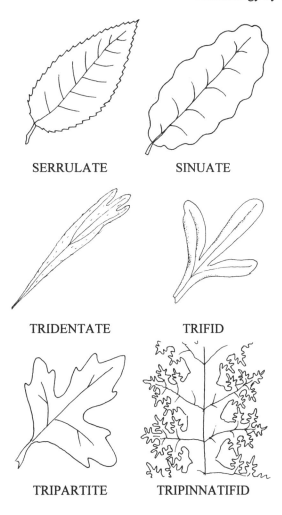

SERRULATE   SINUATE

TRIDENTATE   TRIFID

TRIPARTITE   TRIPINNATIFID

Figure 1593e.

LEAF ATTACHMENT

Amplexicaul. Clasping the stem, as the base or stipules of some leaves. Figure 1594.

Auriculate-clasping. Earlike lobes at the base of a leaf, encircling the stem. Figure 1595.

Clasping. Wholly or partly surrounding the stem. Figure 1594.

Connate-perfoliate. With the bases of opposite leaves fused around the stem. Figure 1596.

Decurrent. Extending downward from the point of insertion, as a leaf base that extends down along the stem. Figure 1597.

Excurrent. Extending beyond what is typical, as

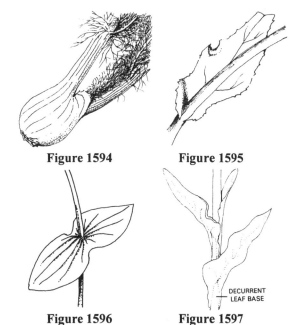

Figure 1594   Figure 1595

Figure 1596   Figure 1597

DECURRENT LEAF BASE

in a leaf base which extends down the stem. See illustration for **decurrent**.

Ocreate. With sheathing stipules. Figure 1598.

Perfoliate. A leaf with the margins entirely surrounding the stem, so that the stem appears to pass through the leaf. Figure 1599.

Petiolate. With a petiole. Figure 1600.

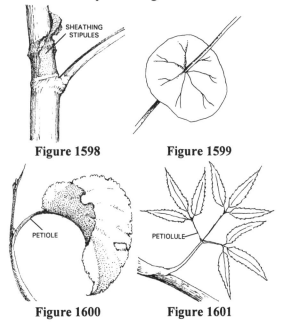

SHEATHING STIPULES

Figure 1598   Figure 1599

PETIOLE   PETIOLULE

Figure 1600   Figure 1601

Petioled. See **petiolate**.

Petiolulate. With a petiolule. Figure 1601.

Sessile. Attached directly, without a supporting stalk, as a leaf without a petiole. Figure 1602.

Sheathing. Forming a sheath, as the leaf base of a grass forms a sheath as it surrounds the stem. Figure 1603.

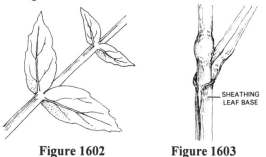

| Figure 1602 | Figure 1603 |

LEAF ARRANGEMENT   (Figure 1604.)

Key to Common Leaf Arrangements

1 Leaves positioned at the base of the stem. . **Basal**
1 Leaves positioned on an elongated stem.
    (**Cauline**)
    2 Leaves one per node. . . . . **Alternate**
    2 Leaves two or more per node.
      3 Leaves two per node. . . . . **Opposite**
      3 Leaves three or more per node. . **Whorled**

Alternate. Borne singly at each node, as leaves on a stem. (compare **opposite**)

Basal. Positioned at or arising from the base, as leaves arising from the base of the stem.

Bilateral. Arranged on two sides, as leaves on a stem.

Cauline. Leaves arising from the stem above ground level.

Decussate. Arranged along the stem in pairs, with each pair at right angles to the pair above or below.

Dextrorse. Turned to the right or spirally arranged to the right, as in the leaves on some stems.

Distichous. In two vertical ranks or rows on opposite sides of an axis.

Equitant. Overlapping or straddling in two ranks, as the leaves of *Iris*.

Opposite. Borne across from one another at the same node, as in a stem with two leaves per node. (compare **alternate**)

Ranked. Arranged into vertical rows.

Rosette. A dense radiating cluster of leaves usually at or near ground level.

Rosulate. With the leaves arranged in basal rosettes, the stem very short or lacking.

Sinistrorse. Turned to the left or spirally arranged to the left, as in the leaves on some stems.

Verticillate. Arranged in verticils; whorled.

Whorled. With parts arranged in whorls, as in a leaf arrangement with three or more leaves arising from a node. Same as **verticillate**.

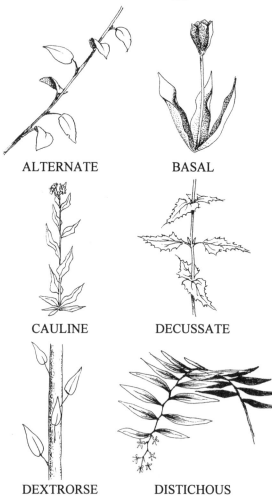

Figure 1604a.

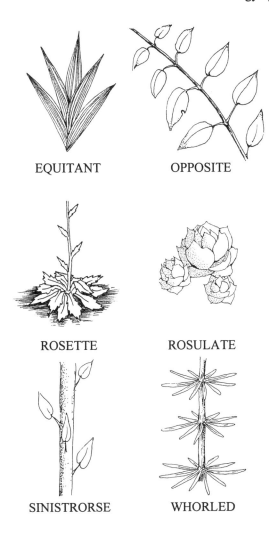

EQUITANT   OPPOSITE

ROSETTE   ROSULATE

SINISTRORSE   WHORLED

Figure 1604b.

MISCELLANEOUS LEAF TERMS

Circinate. Coiled from the tip downward, as in the young leaves of a fern. Figure 1605.

Cirriferous. Bearing a tendril. Figure 1606.

Complicate. Folded together. Figure 1607.

Conduplicate. Folded together lengthwise with the upper surface within, as the leaves of

Figure 1605

many grasses. Figure 1608.

Deciduous. Falling off, as leaves from a tree; not evergreen; not persistent.

Estipulate. Without stipules. Figure 1609.

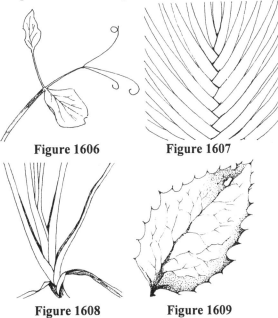

Figure 1606   **Figure 1607**

Figure 1608   **Figure 1609**

Evergreen. Having green leaves through the winter; not deciduous.

Exstipulate. Same as **estipulate**.

Frond. A large, divided leaf; a fern or palm leaf. Figure 1610.

Gamophyllous. With the leaves united, usually by the margins.

Heterophyllous. With different kinds of leaves on the same plant. Figure 1611.

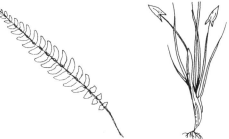

Figure 1610   **Figure 1611**

Limbate. Bordered, as in a leaf or flower in which one color forms an edging or margin around another. Figure 1612.

Macrophyll. The relatively large, expanded leaf of

higher vascular plants. Figure 1613.

Marcescent. Withering but persistent, as the sepals and petals in some flowers or the leaves at the base of some plants. Figure 1614.

Marginate. With a distinct margin.

Phyllopodic. With the lowest leaves well developed, not reduced to scales. Figure 1615.

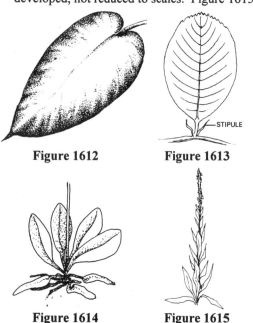

Figure 1612　　　　　**Figure 1613**

Figure 1614　　　　　**Figure 1615**

Stipulate. Bearing stipules. Figure 1613.

Succulent. Juicy and fleshy, as the leaves of *Aloe*. Figure 1616.

Venation. The pattern of veining on a leaf. Figure 1617.

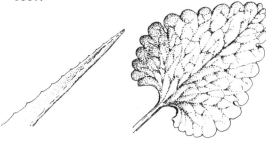

Figure 1616　　　　　**Figure 1617**

SURFACES

Surfaces of leaves, stems, fruits, and other organs.

Aculeate. Prickly; covered with prickles. Figure 1618.

Aculeolate. Minutely prickly; covered with tiny prickles. Figure 1619.

Figure 1618　　　　　**Figure 1619**

Alveolar. See **alveolate**.

Alveolate. Honey-combed, with pits separated by thin, ridged partitions. Figure 1620.

Aperturate. With one or more openings or apertures. In pollen grains, these apertures may be only thin spots rather than actual perforations. Figure 1621.

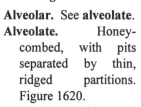

Figure 1620

Arachnoid. Bearing long, cobwebby, entangled hairs. Figure 1622.

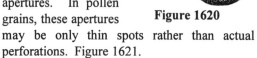

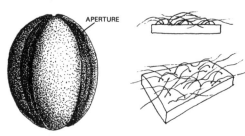

Figure 1621　　　　　**Figure 1622**

Argenteous. Silvery.

Armed. Bearing thorns, spines, barbs, or prickles.

Asperous. Rough to the touch.

Barbellate. With short, stiff hairs or barbs. Figure 1623.

Barbellulate. With very tiny short, stiff hairs or barbs. Figure 1624.

Bullate. With rounded, blistery projections covering the surface. Figure 1625.

Canaliculate. With longitudinal channels or grooves. Figure 1626.

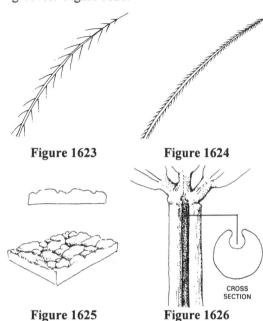

Figure 1623   **Figure 1624**

Figure 1625   **Figure 1626**

CROSS SECTION

Cancellate. Latticed with a fine, regular, reticulate pattern. Figure 1627.

Canescent. Gray or white in color due to a covering of short, fine gray or white hairs. Figure 1628.

Figure 1627   **Figure 1628**

Channeled. With one or more deep longitudinal grooves. Figure 1629.

Ciliate. With a marginal fringe of hairs. Figure 1630.

Ciliolate. With a marginal fringe of minute hairs. Figure 1631.

Figure 1629

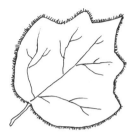

Figure 1630   **Figure 1631**

Cinereous. Ash-colored; grayish due to a covering of short hairs.

Coriaceous. With a leathery texture.

Corrugated. Wrinkled or folded into alternating furrows and ridges. Figure 1632.

Crinite. With tufts of long, soft hairs. Figure 1633.

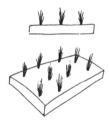

Figure 1632   **Figure 1633**

Echinate. With prickles or spines. Figure 1634.

Echinulate. With very small prickles or spines. Figure 1635.

Figure 1634   **Figure 1635**

Faceted. With many plane surfaces, like a cut gem, as in some seeds. Figure 1636.

Farinose. Covered with a mealy, powdery substance. Figure 1637.

Faveolate. Honeycombed or pitted; alveolate.

Favose. See **faveolate**.

Fenestrate. With window-like perforations, open-

Key to Common Surfaces

1 Surface lacking hairs, grooves, scales, glands, pits, perforations, or projections; smooth.
  2 Surface sticky to the touch. . . . . **Viscid, Glutinous**
  2 Surface not obviously sticky to the touch.
    3 Surface covered with a whitish or bluish waxy coating. . . . . **Glaucous, Pruinose**
    3 Surface lacking a whitish or bluish waxy coating.
      4 Surface covered with a mealy, powdery substance. . . . . **Farinose**
      4 Surface not covered with a mealy, powdery substance. . . . . **Glabrous**
1 Surface with hairs, grooves, scales, glands, pits, perforations, or projections.
  5 Surface with hairs.
    6 Hairs limited to the margins of the surface. . . . . **Ciliate, Fimbriate, Fringed**
    6 Hairs not limited to the margins of the surface.
      7 Hairs not evenly spread over the surface, occurring in tufts. . . . . **Floccose, Crinite**
      7 Hairs evenly spread over the surface.
        8 Hairs with hooks or barbs.
          9 Hairs hooked at the tip. . . . . **Uncinate**
          9 Hairs barbed.
            10 Hairs barbed from apex to base. . . . . **Barbellate**
            10 Hairs barbed only near the apex. . . . . **Glochidiate**
        8 Hairs lacking hooks or barbs.
          11 Hairs bearing glands. . . . . **Glandular**
          11 Hairs lacking glands.
            12 Hairs with several branches radiating from a central point. . . . . **Stellate**
            12 Hairs simple or forked, but not with several branches radiating from a central point.
              13 Hairs straight, not interwoven or entangled.
                14 Hairs stiff and sharp.
                  15 Hairs borne on swollen, nipple-like bases. . . . . **Papillose-hispid**
                  15 Hairs not borne on swollen, nipple-like bases.
                    16 Hairs appressed. . . . . **Strigose**
                    16 Hairs not appressed.
                      17 Hairs very short. . . . . **Scabrous**
                      17 Hairs longer.
                        18 Hairs stiff enough to break the skin. . . . . **Hirsute**
                        18 Hairs not stiff enough to break the skin. . . . . **Hispid**
                14 Hairs soft and flexible.
                  19 Hairs long.
                    20 Hairs appressed. . . . . **Sericeous**
                    20 Hairs not appressed. . . . . **Pilose**
                  19 Hairs short.
                    21 Hairs very short and dense, producing a whitish appearance. . . . . **Canescent**
                    21 Hairs somewhat longer, not producing a whitish appearance. . . . . **Pubescent**
              13 Hairs curly or wavy and usually interwoven or entangled.
                22 Hairs long.
                  23 Hairs dense, so that the leaf surface is obscured. . . . . **Lanate, Woolly**
                  23 Hairs less dense, the leaf surface not obscured. . . . . **Villous, Holosericeous**
                22 Hairs short.
                  24 Hairs matted. . . . . **Tomentose**
                  24 Hairs not matted. . . . . **Velutinous**

ings, or translucent areas. Figure 1638.

Fimbriate. Fringed, usually with hairs or hair-like structures (fimbrillae) along the margin. Figure 1639.

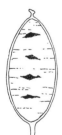

Figure 1636   **Figure 1637**

Figure 1638   **Figure 1639**

Floccose. Bearing tufts of long, soft, tangled hairs. Figure 1640.

Flocculent. Bearing tufts of very fine woolly hairs; floccose.

Fluted. With furrows or grooves. Figure 1641.

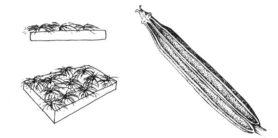

Figure 1640   **Figure 1641**

Foveate. With foveae; pitted. Figure 1642.

Foveolate. With foveolae; minutely pitted. Figure 1643.

Figure 1642   **Figure 1643**

Fringed. With hairs or bristles along the margin.

Furfuraceous. Scurfy; branlike; flaky. Figure 1644.

Glabrate. Becoming glabrous; almost glabrous.

Glabrescent. See **glabrate**.

Glabrous. Smooth; hairless.

Glandular. Bearing glands. Figure 1645.

white short, fine hairs. Figure 1651.

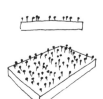

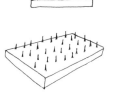

Figure 1644   **Figure 1645**

Figure 1650   **Figure 1651**

Glaucescent. Somewhat glaucous; becoming glaucous.

Glaucous. Covered with a whitish or bluish waxy coating (bloom), as on the surface of a plum.

Glochidiate. Barbed at the tip. Figure 1646.

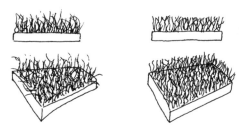
Figure 1646 illustration is at right.

Figure 1646

Glutinous. Gluey; sticky; gummy; covered with a sticky exudation.

Hirsute. Pubescent with coarse, stiff hairs. Figure 1647.

Hirsutulous. Pubescent with very small, coarse, stiff hairs. Figure 1648.

Holosericeous. Covered with fine, silky hairs.

Incanous. With a whitish pubescence.

Inermous. See **unarmed**.

Innocuous. Harmless; lacking thorns or spines.

Laevigate. Lustrous; shining.

Lanate. Woolly; densely covered with long tangled hairs. Figure 1652.

Lanuginose. See **lanuginous**.

Lanuginous. Downy or woolly; with soft downy hairs. Figure 1653.

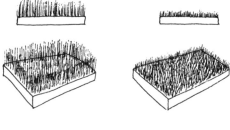

Figure 1647   **Figure 1648**

Figure 1652   **Figure 1653**

Lanulose. Diminutive of lanate; minutely woolly. Figure 1654.

Lepidote. Covered with small, scurfy scales. Figure 1655.

Hirtellate. Same as **hirsutulous**.

Hirtellous. Same as **hirsutulous**.

Hispid. Rough with firm, stiff hairs. Figure 1649.

Hispidulous. Minutely hispid. Figure 1650.

Hoary. With gray or

Figure 1649

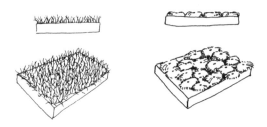

Figure 1654   **Figure 1655**

Lucid. Luminous; shining.

Lustrous. Shiny or glossy.

Mammillate. With nipple-like protuberances. Figure 1656.

Manicate. With a thick, interwoven pubescence. Figure 1657.

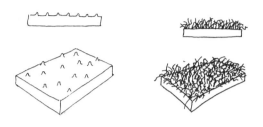

Figure 1656   **Figure 1657**

Mealy. With the consistency of meal; powdery, dry, and crumbly. Figure 1658.

Muricate. Rough with small, sharp projections or points. Figure 1659.

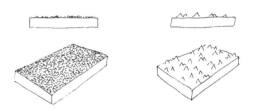

Figure 1658   **Figure 1659**

Muriculate. Very finely muricate. Figure 1660.

Nacreous. With a pearly luster; pearlescent.

Nitid. Lustrous; shining.

Notate. Marked with lines or spots. Figure 1661.

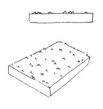

Figure 1660   **Figure 1661**

Paleaceous. Chaffy; with chaffy scales. Figure 1662.

Pannose. Covered with a short, dense, felt-like tomentum. Figure 1663.

Papillate. Having papillae. Figure 1664.

Papillose. Having minute papillae. Figure 1665.

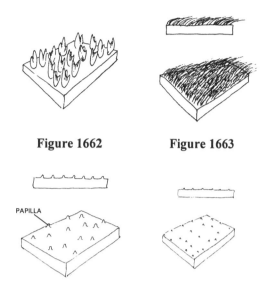

Figure 1662   **Figure 1663**

PAPILLA

Figure 1664   **Figure 1665**

Papillose-hispid. With stiff hairs borne on swollen, nipple-like bases. Figure 1666.

Papyraceous. Papery in texture and usually color.

Pellucid. Transparent or translucent.

Perforate. With holes or perforations. Figure 1667.

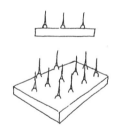

Figure 1666   **Figure 1667**

Pilose. Bearing long, soft, straight hairs. Figure 1668.

Pilosulose. Bearing minute, long, soft, straight hairs. Figure 1669.

Pilosulous. See **pilosulose**.

Pitted. With small pits or depressions. Figure 1670.

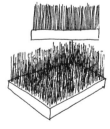

Figure 1668

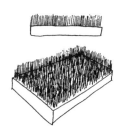

Figure 1669 **Figure 1670**

Plumbeous. Lead-colored.

Pruinate. See **pruinose**.

Pruinose. With a waxy, powdery, usually whitish coating (bloom) on the surface; conspicuously glaucous, like a prune.

Puberulent. Minutely pubescent; with fine, short hairs. Figure 1671.

Puberulous. See **puberulent**.

Pubescent. Covered with short, soft hairs; bearing any kind of hairs. Figure 1672.

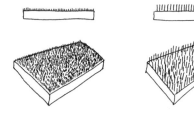

Figure 1671 **Figure 1672**

Pulverulent. Appearing dusty or powdery.

Punctate. Dotted with pits or with translucent, sunken glands or with colored dots. Figure 1673.

Puncticulate. Minutely punctate. Figure 1674.

Figure 1673 **Figure 1674**

Pustular. See **pustulose**.

Pustulate. See **pustulose**.

Pustuliferous. See **pustulose**.

Pustulose. With small blisters or pustules, often at the base of a hair. Figure 1675.

Ramentaceous. With flattened, scaly outgrowths, as on the epidermis of the stem and leaves of some ferns. Figure 1676.

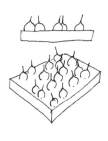

Figure 1675 **Figure 1676**

Roridulate. With a covering of waxy platelets, appearing moist.

Rugate. See **rugose**.

Rugose. Wrinkled. Figure 1677.

Rugulose. Slightly wrinkled. Figure 1678.

Ruminate. Roughly wrinkled, as if chewed. Figure 1679.

Figure 1677

Figure 1678 **Figure 1679**

Scaberulent. See **scaberulose**.

Scaberulose. Slightly rough to the touch, due to the structure of the epidermal cells, or to the presence of short stiff hairs. Figure 1680.

Scaberulous. See **scaberulose**.

Scabrellate. Same as **scaberulose**.

Scabrid. Roughened.

Scabridulous. Minutely roughened.

Scabrous. Rough to the touch, due to the structure

of the epidermal cells, or to the presence of short stiff hairs. Figure 1681.

Figure 1680   **Figure 1681**

Scurfy. Covered with small, bran-like scales. Figure 1682.

Sericeous. Silky, with long, soft, slender, somewhat appressed hairs. Figure 1683.

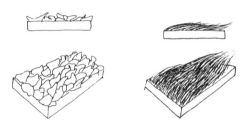

Figure 1682   **Figure 1683**

Setose. Covered with bristles. Figure 1684.

Setulose. Covered with minute bristles. Figure 1685.

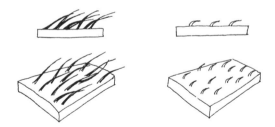

Figure 1684   **Figure 1685**

Silky. Silk-like in appearance or texture; sericeous.

Smooth. With an even surface; not rough to the touch.

Spiniferous. See **spinose**.

Spinose. Bearing spines.

Spinous. See **spinose**.

Spinulose. Bearing spinules.

Spiny. With spines.

Squamate. Covered with scales (squamae). See illustration for **scurfy**.

Squamulose. With minute squamellae.

Stellate. Star-shaped, as in hairs with several to many branches radiating from the base. Figure 1686.

Striate. Marked with fine, usually parallel lines or grooves. Figure 1687.

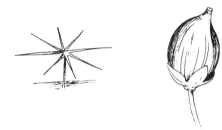

Figure 1686   **Figure 1687**

Strigillose. Minutely strigose. Figure 1688.

Strigose. Bearing straight, stiff, sharp, appressed hairs. Figure 1689.

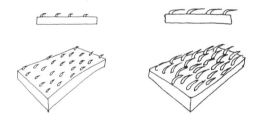

Figure 1688   **Figure 1689**

Strigulose. See **strigillose**.

Strumose. With a covering of cushion-like swellings; bullate. Figure 1690.

Sulcate. With longitudinal grooves or furrows. Figure 1691.

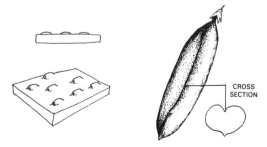

CROSS SECTION

Figure 1690   **Figure 1691**

Tesselate. With a checkered pattern. Figure 1692.
Tomentellous. See **tomentulose.**
Tomentose. With a covering of short, matted or tangled, soft, wooly hairs; with tomentum. Figure 1693.

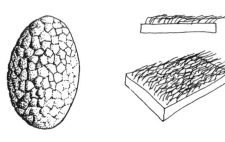

Figure 1692   **Figure 1693**

Tomentulose. Slightly tomentose. Figure 1694.
Translucent. Almost transparent. Figure 1695.

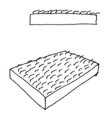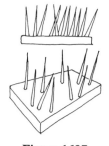

Figure 1694   **Figure 1695**

Unarmed. Lacking spines, prickles, or thorns.
Uncinate. Hooked at the tip. Figure 1696.
Urent. Stinging. Figure 1697.

Figure 1696   **Figure 1697**

Velutinous. Velvety; covered with short, soft, spreading hairs. Figure 1698.
Verrucose. Warty; covered with wart-like elevations. Figure 1699.
Villose. Same as **villous.**
Villosulous. Diminutive if **villous.**
Villous. Bearing long, soft, shaggy, but unmatted,

hairs. Figure 1700.
Viscid. Sticky or gummy.
Viscidulous. Slightly sticky.
Woolly. With long, soft, entangled hairs; lanate. Figure 1701.

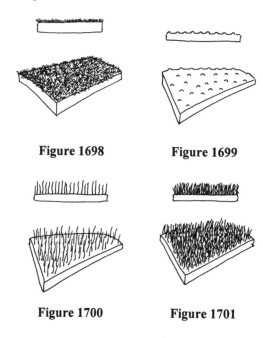

Figure 1698   **Figure 1699**

Figure 1700   **Figure 1701**

INFLORESCENCES

The flowering part of a plant; a flower cluster; the arrangement of the flowers on the flowering axis.

INFLORESCENCE PARTS

Bract. A reduced leaf or leaf-like structure at the base of a flower or inflorescence. Figure 1702.
Bracteole. A small bract borne on a peduncle. Figure 1703.

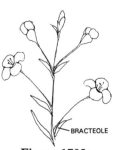

EPICALYX BRACT   BRACTEOLE

Figure 1702   **Figure 1703**

Bractlet. See **bracteole**.

Cupule. A cup-shaped involucre, as in an acorn. Figure 1704.

Disk. In the Compositae (Asteraceae), the central portion of the involucrate head bearing tubular or disk flowers. Figure 1705.

Figure 1704   **Figure 1705**

Epicalyx. An involucre which resembles an outer calyx, as in *Malva*. Figure 1702.

Floret. A small flower; an individual flower within a dense cluster, as a grass flower in a spikelet, or a flower of the Compositae (Asteraceae) in an involucrate head. Figures 1706 and 1707.

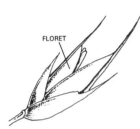

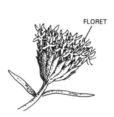

Figure 1706   **Figure 1707**

Flower. The reproductive portion of the plant, consisting of stamens, pistils, or both, and usually including a perianth of sepals or both sepals and petals. Figure 1708.

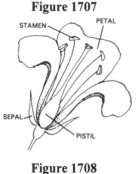

Figure 1708

Involucel. A small involucre; a secondary involucre, as in the bracts of the secondary umbels in the Umbelliferae (Apiaceae). Figure 1709.

Involucre. A whorl of bracts subtending a flower or flower cluster. Figure 1710.

Figure 1709   **Figure 1710**

Involucrum (pl. **involucra**). See **involucre**.

Ocreola (pl. **ocreolae**). A minute stipular sheath around the secondary divisions of the inflorescence in some members of the Polygonaceae.

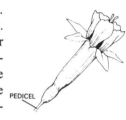

Pedicel. The stalk of a single flower in an inflorescence, or of a grass spikelet. Figure 1711.

Figure 1711

Peduncle. The stalk of a solitary flower or of an inflorescence. Figures 1712 and 1713.

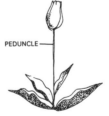

Figure 1712   **Figure 1713**

Perigynium (pl. **perigynia**). A scale-like bract enclosing the pistil in *Carex*. Figure 1714.

Phyllary. An involucral bract of the Compositae (Asteraceae). Figure 1715.

Rachilla. The axis of a grass or sedge spikelet; a small rachis. Figure 1716.

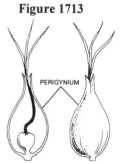

Figure 1714

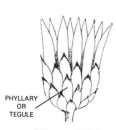

Figure 1715   **Figure 1716**

Rachis. The main axis of an inflorescence. Figure 1717.

Ray. An inflorescence branch in an umbel. Figure 1718.

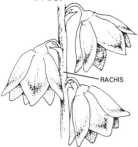

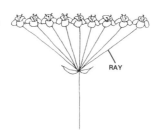

Figure 1717   **Figure 1718**

Scape. A leafless peduncle arising from ground level (usually from a basal rosette) in acaulescent plants. Figure 1719.

Secondary peduncle. An inflorescence branch. Figure 1720.

Spathe. A large bract or pair of bracts subtending and often enclosing an inflorescence. Figure 1721.

Tegule. One of the bracts of the involucre in the Compositae (Asteraceae). Figure 1715.

Figure 1719

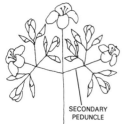

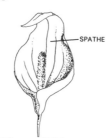

Figure 1720   **Figure 1721**

INFLORESCENCE TYPES   (Figure 1722.)

Ament. See **catkin**.

Capitulum. A small flower head.

Catkin. An inflorescence consisting of a dense spike or raceme of apetalous, unisexual flowers as in Salicaceae and Betulaceae; an ament.

Cincinnus. A dense helicoid cyme with the pedicels short on the developed side.

Corymb. A flat-topped or round-topped inflorescence, racemose, but with the lower pedicels longer than the upper.

Cyathium (pl. **cyathia**). The inflorescence in the genus *Euphorbia*, consisting of a cup-like involucre containing a single pistil and male flowers with a single stamen.

Cyme. A flat-topped or round-topped determinate inflorescence, paniculate, in which the terminal flower blooms first.

Cymule. A small cyme or a small section of a compound cyme.

Dichasium. A cymose inflorescence in which each axis produces two opposite or subopposite lateral axes.

Glomerule. A dense cluster; a dense, head-like cyme.

Head. A dense cluster of sessile or subsessile flowers; the involucrate inflorescence of the Compositae (Asteraceae).

Helicoid cyme. A one-sided cymose inflorescence coiled like a spiral or helix.

Hypanthodium. An inflorescence with flowers borne on the walls of a capitulum, as in *Ficus*.

Intercalary inflorescence. An inflorescence type in which the main vegetative axis of the plant continues to elongate after the flowers are produced.

Mixed inflorescence. An inflorescence with both racemose and cymose portions.

Monochasium. A type of cymose inflorescence with only a single main axis.

Panicle. A branched, racemose inflorescence with flowers maturing from the bottom upwards.

Pleiochasium. A cymose inflorescence with more than two branches from the main axis.

Polychasium. A cymose inflorescence in which each axis produces more than two lateral axes.

Key to Common Inflorescence Types

1 Flowers sessile.
  2 Inflorescence elongate.
    3 Flowers very small and densely clustered, obscuring the inflorescence axis.
      4 Perianth reduced to paired bracts. . . . . **Spikelet**
      4 Perianth not reduced to paired bracts.
        5 Inflorescence usually erect, bisexual, with a thickened axis. **Spadix**
        5 Inflorescence usually pendulous, unisexual, lacking a thickened axis. . . . . **Catkin, Ament**
    3 Flowers not very small and densely clustered, not obscuring the inflorescence axis.
      6 Perianth reduced to paired bracts. . . . . **Spikelet**
      6 Perianth not reduced to paired bracts, often showy. . . . . **Spike**
  2 Inflorescence not elongate.
    7 Flowers enclosed within the walls of a concave capitulum. **Hypanthodium**
    7 Flowers borne on a flat or convex receptacle. . . . . **Head, Capitulum**
1 Flowers pedicellate.
  8 Inflorescence unbranched.
    9 Pedicels arising from a common point, like the struts of an umbrella. . . . . **Umbel**
    9 Pedicels not arising from a common point.
     10 Inflorescence determinate, the central or terminal flower developing first. . . . . **Cyme**
     10 Inflorescence indeterminate, the lateral or basal flowers developing first.
      11 Inflorescence flat-topped or rounded. . . . . **Corymb**
      11 Inflorescence elongate. . . . . **Raceme**
  8 Inflorescence branched.
    12 Inflorescence flat-topped or rounded.
     13 Inflorescence branches arising from a common point, like the struts of an umbrella. . . . .
      . . . . **Compound umbel**
     13 Inflorescence branches not arising from a common point.
      14 Inflorescence determinate, the central or terminal flower developing first. . . . . . . **Compound cyme**
      14 Inflorescence indeterminate, the lateral or basal flowers developing first. . . . . . **Compound corymb**
    12 Inflorescence elongate.
     15 Flowers densely clustered in a compact, cylindrical or ovate inflorescence. . . . . **Thyrse**
     15 Flowers less densely clustered in a more open inflorescence. . . . . **Panicle**

Pseudanthium. A compact inflorescence of many small flowers which simulates a single flower.

Raceme. An unbranched, elongated inflorescence with pedicellate flowers maturing from the bottom upwards.

Scorpioid cyme. A determinate cymose inflorescence with a zigzag rachis.

Solitary. Flowers occurring singly and not borne in a cluster or group.

Spadix. A spike with small flowers crowded on a thickened axis.

Spike. An unbranched, elongated inflorescence with sessile or subsessile flowers or spikelets maturing from the bottom upwards.

Spikelet. A small spike or secondary spike; the ultimate flower cluster of grasses and sedges, consisting of one to many flowers subtended by two bracts (glumes).

Strobile. A cone or an inflorescence resembling a cone.

Thyrse. A compact, cylindrical, or ovate panicle with an indeterminate main axis and cymose sub-axes.

Umbel. A flat-topped or convex inflorescence with the pedicels arising more or less from a common point, like the struts of an umbrella; a highly condensed raceme.

Umbellet. An ultimate umbellate cluster of a compound umbel.

Verticillaster. A pair of axillary cymes arising from opposite leaves or bracts and forming a false whorl.

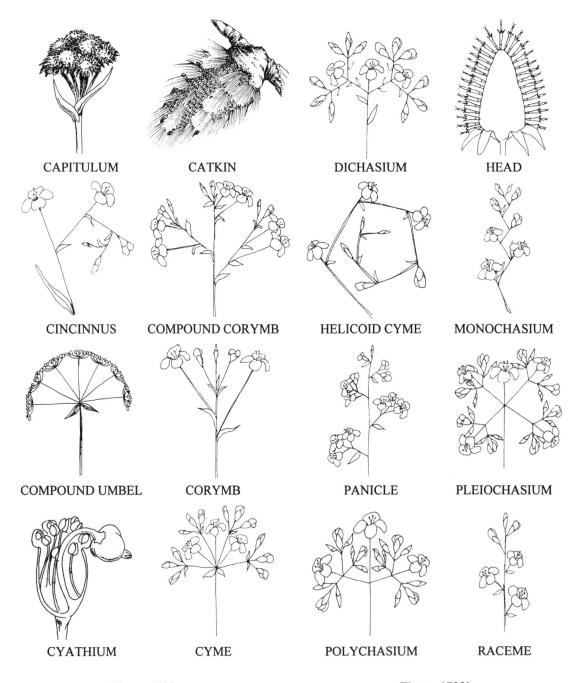

CAPITULUM   CATKIN   DICHASIUM   HEAD

CINCINNUS   COMPOUND CORYMB   HELICOID CYME   MONOCHASIUM

COMPOUND UMBEL   CORYMB   PANICLE   PLEIOCHASIUM

CYATHIUM   CYME   POLYCHASIUM   RACEME

Figure 1722a.   **Figure 1722b.**

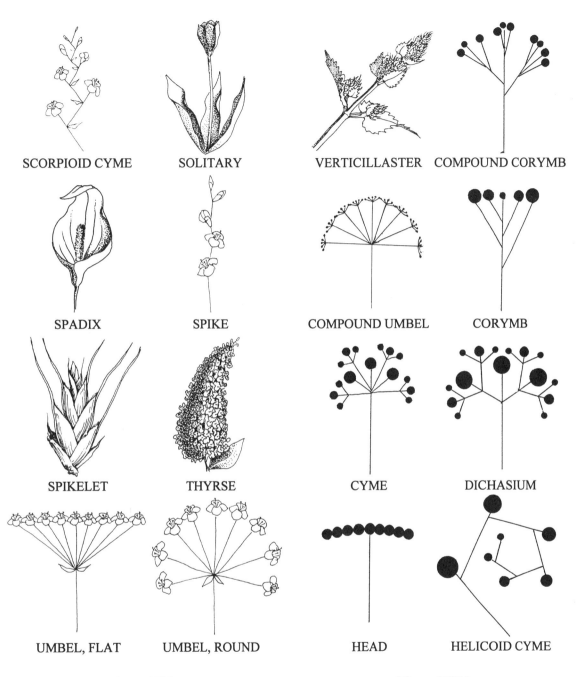

SCORPIOID CYME

SOLITARY

VERTICILLASTER

COMPOUND CORYMB

SPADIX

SPIKE

COMPOUND UMBEL

CORYMB

SPIKELET

THYRSE

CYME

DICHASIUM

UMBEL, FLAT

UMBEL, ROUND

HEAD

HELICOID CYME

Figure 1722c.

Figure 1722d.

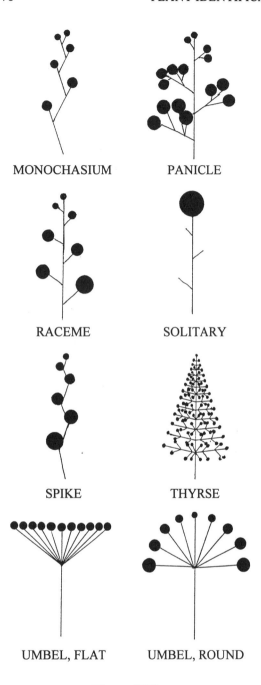

MONOCHASIUM   PANICLE

RACEME   SOLITARY

SPIKE   THYRSE

UMBEL, FLAT   UMBEL, ROUND

Figure 1722e.

INFLORESCENCE FORMS

Androgynous. An inflorescence with both staminate and pistillate flowers, the staminate flowers borne above the pistillate, as in some *Carex* species. (compare **gynaecandrous**)

Axillary. Positioned in or arising in an axil. Figure 1723.

Capitate. Head-like, or in a head-shaped cluster, as the flowers in the Compositae (Asteraceae). Figure 1724.

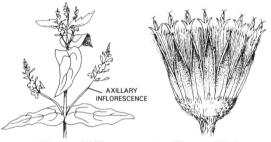

Figure 1723   **Figure 1724**

Capitellate. With small head-like structures, or with parts in very small head-shaped clusters. Figure 1725.

Centrifugal inflorescence. A flower cluster developing from the center outward, as in a cyme. Figure 1726.

Figure 1725   **Figure 1726**

Centripetal inflorescence. A flower cluster developing from the edge toward the center, as in a corymb. Figure 1727.

Corymbiform. An inflorescence with the general appearance, but not necessarily the

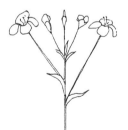

Figure 1727

structure, of a true corymb.

Corymbose. Having flowers in corymbs. The term is sometimes used in the same sense as **corymbiform**. Figure 1727.

Cyathiform. With the form of a cyathium; cup-shaped.

Cymose. With flowers in a cyme. Figure 1726.

Determinate. Describes an inflorescence in which the terminal flower blooms first, halting further elongation of the main axis. Figure 1726.

Glomerate. Densely clustered. Figure 1728.

Glomerulate. Arranged in very small, dense clusters. Figure 1729.

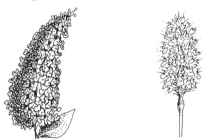

Figure 1728   **Figure 1729**

Gynaecandrous. An inflorescence with the pistillate flowers borne above the staminate, as in some *Carex* species. (compare **androgynous**)

Helicoid. Coiled like a spiral or helix, as in some one-sided cymose inflorescences in the Boraginaceae. Figure 1730.

Indeterminate. Describes an inflorescence in which the lower or outer flowers bloom first, allowing indefinite elongation of the main axis. Figure 1727.

Monochasial. With the form of a monochasium. Figure 1731.

Figure 1730   **Figure 1731**

Paniculate. Having flowers in panicles. Figure 1732.

Paniculiform. An inflorescence with the general appearance, but not necessarily the structure, of a true panicle.

Racemiform. An inflorescence with the general appearance, but not necessarily the structure, of a true raceme.

Racemose. Having flowers in racemes. The term is sometimes used in the same sense as **racemiform**. Figure 1733.

Figure 1732   **Figure 1733**

Scorpioid. A determinate cymose inflorescence with a zigzag rachis. Figure 1734; often used in the same sense as **helicoid**.

Secund. Arranged on one side of the axis only. Figure 1735.

Figure 1734   **Figure 1735**

Spathaceous. Spathe bearing; spathe-like. Figure 1736.

Spicate. Arranged in a spike. Figure 1737.

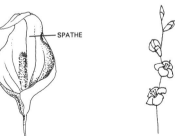

Figure 1736   **Figure 1737**

Spiciform. An inflorescence with the general appearance, but not necessarily the structure, of a true spike.

Terminal. At the tip or apex. Figure 1738.

Thyrsoid. Thyrse-like.

Umbellate. In umbels; umbel-like. Figure 1739.

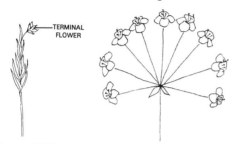

Figure 1738   **Figure 1739**

Umbelliform. An inflorescence with the general appearance, but not necessarily the structure, of a true umbel. The term is often applied to inflorescences which are condensed cymes rather than condensed racemes.

FLOWERS

The reproductive portion of the plant, consisting of stamens, pistils, or both, and usually including a perianth of sepals or both sepals and petals.

FLOWER PARTS

Androecium. All of the stamens in a flower, collectively. Figure 1740.

Calyx (pl. **calyces**, **calyxes**). The outer perianth whorl; collective term for all of the sepals of a flower. Figure 1740.

Carpel. A simple pistil formed from one modified leaf, or that part of a compound pistil formed from one modified leaf; megasporophyll. Figure 1741.

Corolla. The collective name for all of the petals of a flower; the inner perianth whorl. Figure 1740.

Floral envelope. A collective term for the calyx and corolla. (same as **perianth**)

Gynecium. See **gynoecium**.

Gynoecium. All of the carpels or pistils of a flow-

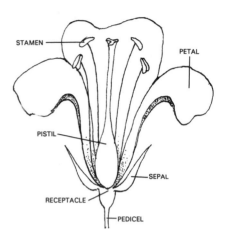

Figure 1740

er, collectively. Figure 1740.

Nectar gland. See **nectary**.

Nectary. A tissue or organ which produces nectar. Figure 1742.

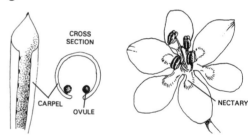

Figure 1741   **Figure 1742**

Pedicel. The stalk of a single flower in an inflorescence. Figure 1740.

Peduncle. The stalk of a solitary flower or of an inflorescence. Figure 1743.

Perianth. The calyx and corolla of a flower, collectively, especially when they are similar in appearance. Figure 1744.

Petal. An individual segment or member of the corolla, usually colored or white. Figure 1740.

Figure 1743

Pistil. The female reproductive organ of a flower, typically consisting of a stigma, style, and ovary. Figure 1740. (compare **gynoecium**)

Receptacle. The portion of the pedicel upon which the flower parts are borne; in the Compositae (Asteraceae), the part of the peduncle where the flowers of the head are borne. Figures 1740 and 1745.

Sepal. A segment of the calyx. Figure 1740.

Stamen (pl. **stamens**, **stamina**). The male reproductive organ of a flower, consisting of an anther and filament; the angiosperm microsporophyll. Figure 1740.

Tepal. A segment of a perianth which is not differentiated into calyx and corolla; a sepal or petal. Figure 1744.

Torus (pl. **tori**). See **receptacle**.

Whorl. A ring-like arrangement of similar parts arising from a common point or node; a verticil. Figure 1746.

FLOWER SYMMETRY

Actinomorphic. Radially symmetrical, so that a line drawn through the middle of the structure along any plane will produce a mirror image on either side. Figure 1747.

Actinomorphous. See **actinomorphic**.

Irregular. Bilaterally symmetrical; said of a flower in which all parts are not similar in size and arrangement on the receptacle. Figure 1748.

Monosymmetrical. Bilaterally symmetrical;

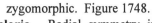

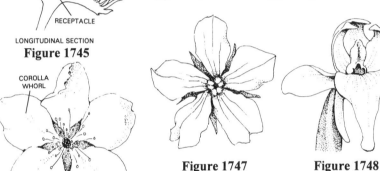

TEPAL

Figure 1744

RECEPTACLE

LONGITUDINAL SECTION

Figure 1745

COROLLA WHORL

Figure 1746

zygomorphic. Figure 1748.

Peloria. Radial symmetry in flowers normally bilaterally symmetrical.

Regular. Radially symmetrical; said of a flower in which all parts are similar in size and arrangement on the receptacle. Figure 1747.

Stereomorphic. Radially symmetrical, so that a line drawn through the middle of the structure along any plane will produce a mirror image on either side; essentially the same as **actinomorphic**. Figure 1747.

Zygomorphic. Bilaterally symmetrical, so that a line drawn through the middle of the structure along only one plane will produce a mirror image on either side. Figure 1748.

Zygomorphous. See **zygomorphic**.

Figure 1747   **Figure 1748**

INSERTION OF FLORAL STRUCTURES

Epigynous. With stamens, petals, and sepals attached to the top of the ovary, the ovary inferior to the other floral parts. Figure 1749.

Hypogynous. With stamens, petals, and sepals attached below the ovary, the ovary superior to the other floral parts. Figure 1750.

Perigynous. With stamens, petals, and sepals borne on a calyx tube (hypanthium) surrounding, but not actually attached to, the superior ovary. Figure 1751.

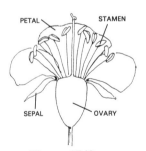

PETAL   STAMEN

SEPAL   OVARY

Figure 1749

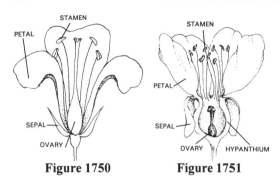

Figure 1750   **Figure 1751**

POLLINATION SYSTEMS

Anemophilous. Wind pollinated; producing wind-borne pollen.

Chasmogamous. Applied to flowers which open before fertilization and are usually cross-pollinated. (compare **cleistogamous**)

Cleistogamous. Said of flowers which self-fertilize without opening. (compare **chasmogamous**)

Dichogamic. See **dichogamous**.

Dichogamous. With the pistils and stamens maturing at different times to prevent self-fertilization. (compare **homogamous**)

Entomophilous. Insect pollinated.

Geitonogamous. Pollinated by flowers of the same plant.

Homogamous. With the pistils and stamens maturing at the same time, allowing self-fertilization. (compare **dichogamous**)

Kleistogamous. See **cleistogamous**.

Metandrous. With the female flowers maturing before the male flowers; protogynous.

Ornithophilous. Pollinated by birds.

Outcrossing. Transferring pollen from the anthers of the flowers of one plant to the stigma of the flower of another plant.

Protandrous. With the anthers releasing pollen before the stigma is receptive.

Proterandrous. See **protandrous**.

Proterogynous. See **protogynous**.

Protogynous. With the stigma receptive before the anthers release pollen.

Self-pollinating. Transferring pollen from the anthers to the stigma of the same flower or to the stigma of another flower on the same plant.

FLOWER SEXUALITY

Andro-dioecious. With staminate and perfect flowers on separate plants.

Andro-monoecious. See **andro-polygamous**.

Andro-polygamous. With staminate and perfect flowers on the same plant.

Bisexual. With both male and female reproductive organs (stamens and pistils). (same as **perfect**)

Diclinous. With the stamens and pistils in separate flowers; imperfect.

Dioecious. With imperfect flowers, the staminate and pistillate flowers borne on different plants. (compare **monoecious**)

Dioicous. See **dioecious**.

Diecious. See **dioecious**.

Gyno-dioecious. With pistillate and perfect flowers on separate plants.

Hermaphroditic. With pistils and stamens in the same flower; bisexual; monoclinous; perfect.

Heterogamous. With flowers of differing sex.

Imperfect. With either stamens or pistils, but not both; unisexual.

Monecious. See **monoecious**.

Monoclinous. With pistils and stamens in the same flower; perfect.

Monoecious. Flowers imperfect, the staminate and pistillate flowers borne on the same plant. (compare **dioecious**)

Perfect. With both male and female reproductive organs (stamens and pistils); bisexual.

Pistillate. Bearing a pistil or pistils, but lacking stamens. (compare **staminate**)

Polygamo-dioecious. Mostly dioecious, but with some perfect flowers.

Polygamo-monoecious. Mostly monoecious, but with some perfect flowers.

Polygamous. With unisexual and bisexual flowers on the same plant.

Staminate. Bearing stamens but not pistils, as a male flower which does not produce fruit or seeds. (compare **pistillate**)

Triecious. See **trioecious**.

Trioecious. With male, female, and bisexual flowers on different plants.

Unisexual. With either male or female reproductive parts, but not both.

FLOWERING TIME

Diurnal. Occurring or opening in the daytime.

Equinoctial. With flowers that open regularly at a particular hour of the day.

Hibernal. Flowering in the winter.

Matutinal. Opening in the morning.

Nocturnal. Opening at night.

Nyctanthous. Night-flowering.

Nyctigamous. Opening at night.

Precocious. With the flowers developing before the leaves.

Proteranthous. With the flowers developing before the leaves.

Semperflorous. Flowering throughout the year.

Serotinous. Flowering late; with flowers developing after the leaves are fully developed.

Vernal. Flowering in the spring.

Vespertine. Opening in the evening.

NUMBERS OF FLORAL STRUCTURES

Anisomerous. With a different number of parts (usually less) than the other floral whorls, as in a flower with five sepals and petals, but only two stamens. Figure 1752.

Complete. With all of the parts typically belonging to it, as a flower with sepals, petals, stamens, and pistils. Figure 1753.

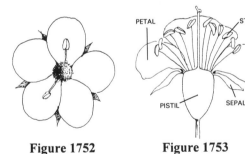

Figure 1752   **Figure 1753**

Dicyclic. With two whorls.

Dimerous. With parts arranged in sets or multiples of two.

Heteromerous. With a variable number of parts, as in a flower with a different number of members in each floral whorl. Figure 1754.

Hexamerous. With parts arranged in sets or multiples of six. Figure 1755.

Figure 1754   **Figure 1755**

Incomplete. Lacking an expected part or series of parts, as in a flower lacking one of the floral whorls (i.e. sepals, petals, stamens, or pistils).

Isomerous. With an equal number of parts, as in a flower with an equal number of members in each floral whorl.

Monocyclic. With a single whorl. Figure 1756.

Monomerous. With a single member, as in a floral whorl with only one part. Figure 1756.

Pentacyclic. With five whorls.

Pentamerous. With parts arranged in sets or multiples of five. Figure 1757.

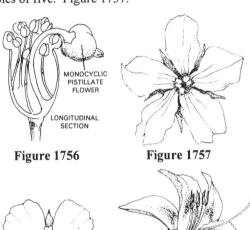

Figure 1756   **Figure 1757**

Figure 1758   **Figure 1759**

Polycyclic. With many whorls.

Polymerous. With many parts, as in a floral whorl with many members.

Symmetric. Said of a flower having the same number of parts in each floral whorl. Figure

1758.

Tetramerous. With parts arranged in sets or multiples of four. Figure 1758.

Trimerous. With parts arranged in sets or multiples of three. Figure 1759.

PERIANTH

Perianth Parts

Ala (pl. **alae**). One of the two lateral petals of a papilionaceous corolla. Figure 1760.

Anterior lip or lobe. The lower lip of a bilabiate corolla. Figure 1761. (compare **posterior**)

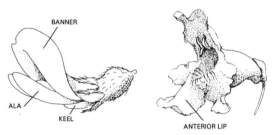

Figure 1760   **Figure 1761**

Banner. The upper and usually largest petal of a papilionaceous flower, as in peas and sweet peas. Figure 1760.

Blade. The broad part of a petal. Figure 1762.

Bridge. A band of tissue connecting the corolla scales, as in *Cuscuta*. Figure 1763.

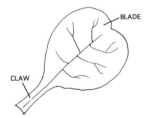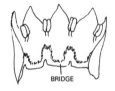

Figure 1762   **Figure 1763**

Calcar. A spur or spurlike appendage. Figure 1764.

Calyx limb. See **calyx lobe**.

Calyx lobe. One of the free portions of a calyx of united sepals. Figure 1765.

Calyx tooth. See **calyx lobe**.

Calyx tube. The tube-like united portion of a calyx of united sepals. Figure 1765.

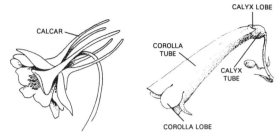

Figure 1764   **Figure 1765**

Carina. A keel or ridge. Figure 1760.

Claw. The narrowed base of some petals and sepals. Figure 1762.

Corolla lobe. One of the free portions of a corolla of united petals. Figure 1765.

Corolla tube. The hollow, cylindric portion of a corolla of united petals. Figure 1765.

Corona. Petal-like or crown-like structures between the petals and stamens in some flowers; a crown. Figure 1766.

Crest. An elevated ridge or rib.

Crown. See **corona**.

Cucullus. A hood. Figure 1767.

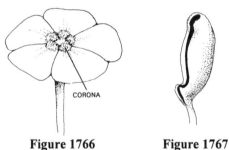

Figure 1766   **Figure 1767**

Epicalyx. An involucre which resembles an outer calyx, as in *Malva*. Figure 1768.

Falls. The sepals of an *Iris*. Figure 1769.

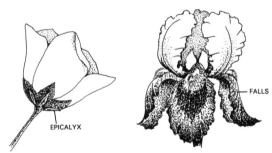

Figure 1768   **Figure 1769**

Floral tube. An elongated tubular portion of a perianth. See illustration for **corolla tube**.

Fornix (pl. **fornices**). One of a set of small crests or scales in the throat of a corolla, as in many of the Boraginaceae. See illustration for **corona**.

Fringe. Hairs or bristles along the margin. Figure 1770.

Galea. The helmet-shaped or hood-like upper lip of some two-lipped corollas. Figure 1771.

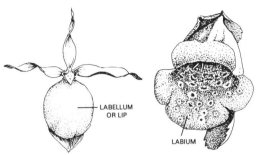

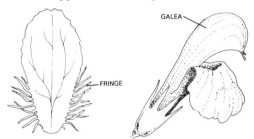

Figure 1775   **Figure 1776**

Lamella (pl. **lamellae**). An erect scale inserted on the petal in some corollas and forming part of the corona. See illustration for **corona**.

Lamina. The expanded portion, or blade, of a petal. Figure 1762.

Lepanthium. A petal with a nectary.

Ligula. See **ligule**.

Ligule. The flattened part of the ray corolla in the Compositae (Asteraceae). Figure 1777.

Limb. The expanded part, or blade, of a petal. Figure 1762; the expanded part of a sympetalous corolla. Figure 1778.

Figure 1770   **Figure 1771**

Gibbosity. A swelling or protuberance; the state of being gibbous. Figure 1772.

Helmet. See **hood**.

Hood. A hollow, arched covering, as the upper petal in *Aconitum*. Figure 1773.

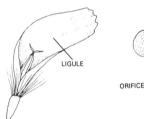

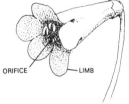

Figure 1777   **Figure 1778**

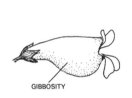

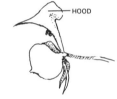

Figure 1772   **Figure 1773**

Horn. A tapering projection resembling the horn of a cow. Figure 1774.

Keel. The two lower united petals of a papillonaceous flower. Figure 1760.

Labellum. Lip; the exceptional petal of an orchid blossom. Figure 1775.

Labium (pl. **labia**). The lower lip of a bilabiate corolla. Figure 1776.

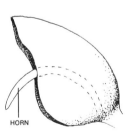

Figure 1774

Lip. One of the two projections or segments of an irregular, two-lipped corolla or calyx; a labium. Figure 1776; the exceptional petal of an orchid blossom. Figure 1775.

Lodicule. Paired, rudimentary scales at the base of the ovary in grass flowers. Figure 1779.

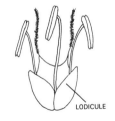

Figure 1779

Orifice. An opening or mouth, as the mouth-like opening of a tubular corolla. Figure 1778.

Palate. A raised appendage on the lower lip of a

corolla which partially or completely closes the throat. Figure 1780.

Pappus. The modified calyx of the Compositae (Asteraceae), consisting of awns, scales, or bristles at the apex of the achene. Figure 1781.

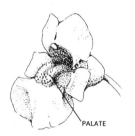

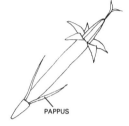

Figure 1780   **Figure 1781**

Petal. An individual segment or member of the corolla, usually colored or white. Figure 1782.

Plait. A fold or pleat, as in some corollas. Figure 1783.

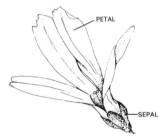

Figure 1782   **Figure 1783**

Posterior. The upper lip of a bilabiate corolla. Figure 1784. (compare **anterior**)

Ray. The strap-like portion of a ligulate flower (or the ligulate flower itself) in the Compositae (Asteraceae). Figure 1785.

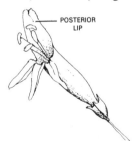

Figure 1784   **Figure 1785**

Sac. A bag-shaped structure. Figure 1786.

Scale. Any thin, flat, scarious structure. Figure 1787.

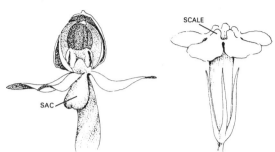

Figure 1786   **Figure 1787**

Sepal. A segment of the calyx. Figure 1782.

Sinus. The cleft, depression, or recess between two lobes of a petal. Figure 1788.

Spur. A hollow, slender, sac-like appendage of a petal or sepal, or of the calyx or corolla. Figure 1789.

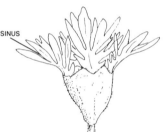

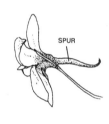

Figure 1788   **Figure 1789**

Squama (pl. **squamae**). A scale, as in some types of pappus in the Compositae (Asteraceae). Figure 1790.

Standard. See **banner**.

Tepal. A segment of a perianth which is not differentiated into calyx and corolla. Figure 1791.

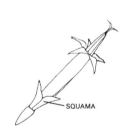

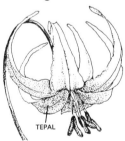

Figure 1790   **Figure 1791**

Throat. The orifice of a gamopetalous corolla or gamosepalous calyx. Figure 1792; the expanded portion of the corolla between the limb and the tube. Figure 1793.

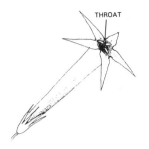

Figure 1792

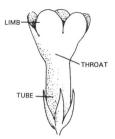

Figure 1793

Vexillum. The upper and usually largest petal of a papilionaceous flower, as in peas and sweet peas; banner. Figure 1794.

Wing. One of the two lateral petals of a papilionaceous corolla. Figure 1794.

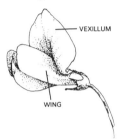

Figure 1794

Perianth Types

Achlamydeous. Lacking a perianth. Figure 1795.
Apetalous. Without petals.
Apopetalous. With separate petals. Figure 1796.
Aposepalous. With separate sepals. Figure 1796.

Figure 1795

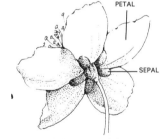

Figure 1796

Asepalous. Without sepals.
Bipetalous. With two petals.
Calyculate. With small bracts around the calyx, as if possessing an outer calyx. Figure 1797.
Chlamydeous. With a floral whorl.
Choripetalous. See **apopetalous** or **polypetalous**.
Dichlamydeous. With two types of perianth whorls, i.e., calyx and corolla. Figure 1796.
Dipetalous. See **bipetalous**.

Double. Having a larger number of petals than usual.
Epappose. Without a pappus. Figure 1798.

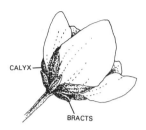

Figure 1797 **Figure 1798**

Gamopetalous. With the petals united, at least partially. Figure 1799.
Gamosepalous. With the sepals united. Figure 1800.

Figure 1799 **Figure 1800**

Monochlamydeous. With only one type of perianth member. Figure 1801.
Monopetalous. See **sympetalous** or **gamopetalous**.
Octopetalous. With eight petals.
Octosepalous. With eight sepals.
Pappiferous. See **pappose**.
Pappose. Pappus-bearing. Figure 1802.
Petaliferous. Bearing petals.
Petalous. With petals.
Polypetalous. With a corolla of completely

Figure 1801

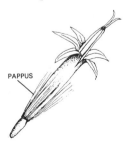

Figure 1802

separate petals. Figure 1803.

Polysepalous. With a calyx of separate sepals. Figure 1803.

Sympetalous. With the petals united, at least near the base. Figure 1804.

Synsepalous. With united sepals. Figure 1804.

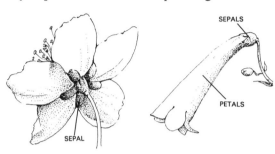

Figure 1803   **Figure 1804**

Tetrapetalous. With four petals.

Tripetalous. With three petals.

Unipetalous. With only a single petal.

Perianth Forms (Figure 1805.)

Accrescent. Becoming larger with age, as a calyx which continues to enlarge after anthesis.

Alate. Winged.

Ampliate. Enlarged or expanded.

Bilabiate. Two-lipped, as in many irregular flowers.

Calcarate. With a calcar; spurred.

Calceolate. Shoe-shaped or slipper-shaped, as the labellum of some orchids.

Campanulate. Bell-shaped.

Carinate. Keeled.

Corniculate. With small horn-like protuberances.

Cornute. Horned. See illustration for **corniculate**.

Coroniform. Crown-shaped.

Cruciate. See **cruciform**.

Cruciform. Cross-shaped.

Cucullate. Hooded or hood-shaped.

Explanate. Spread out flat.

Funnelform. Gradually widening from base to apex; funnel-shaped.

Galeate. Helmet-shaped; with a galea.

Galeiform. See **galeate**.

Gibbous. Swollen or enlarged on one side; ventricose.

Inflated. Swollen or expanded; bladdery.

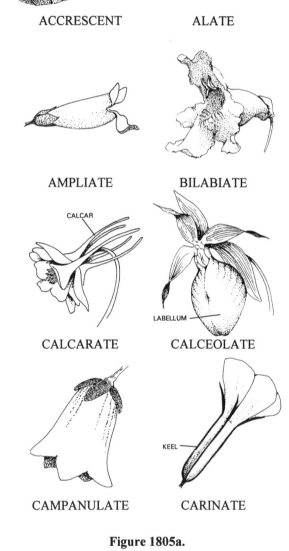

ACCRESCENT   ALATE

AMPLIATE   BILABIATE

CALCARATE   CALCEOLATE

CAMPANULATE   CARINATE

Figure 1805a.

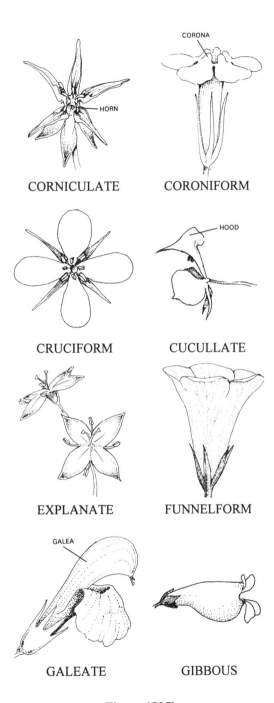

Figure 1805b.

Infundibuliform. Funnel-shaped. See illustration for **funnelform**.

Keeled. Ridged, like the keel of a boat.

Labiate. Lipped; with parts which are arranged like lips or shaped like lips.

Ligulate. Strap-shaped.

Liguliform. See **ligulate**.

Paleolate. With a lodicule.

Papilionaceous. Butterfly-like, as the irregular corolla of a pea, with a banner petal, two wing petals, and two fused keel petals.

Personate. Two-lipped, with the throat closed by a prominent projection (palate).

Plicate. Plaited or folded, as a folding fan.

Porrect. Extended forward; resembling a parrot beak.

Reflexed. Bent backward or downward.

Ringent. Gaping; with widely spreading lips, as in some corollas.

Rotate. Disc-shaped; flat and circular, as a sympetalous corolla with widely spreading lobes and little or no tube.

Saccate. With a sac, or in the shape of a sac.

Salverform. With a slender tube and an abruptly spreading, flattened limb.

Spurred. Bearing a spur or spurs.

Strap-shaped. Elongated and flat.

Tubular. With the form of a tube or cylinder.

Urceolate. Pitcher-like; hollow and contracted near the mouth like a pitcher or urn.

Urn-shaped. See **urceolate**.

Ventricose. Inflated or swollen on one side only, as in some corollas, especially in the genus *Penstemon*.

Winged. Possessing wings.

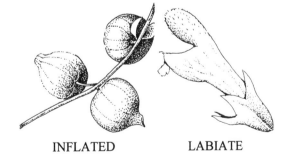

INFLATED   LABIATE

Figure 1805c.

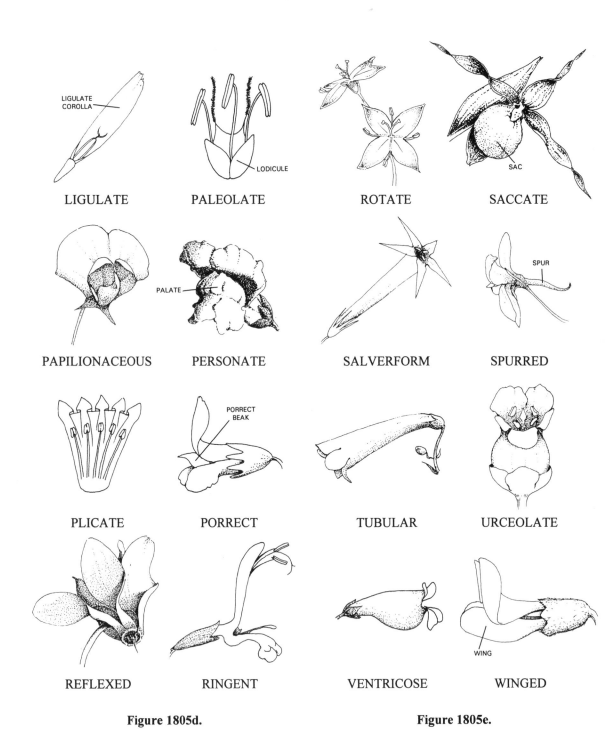

LIGULATE   PALEOLATE   ROTATE   SACCATE

PAPILIONACEOUS   PERSONATE   SALVERFORM   SPURRED

PLICATE   PORRECT   TUBULAR   URCEOLATE

REFLEXED   RINGENT   VENTRICOSE   WINGED

Figure 1805d.   **Figure 1805e.**

ANDROECIUM

The male reproductive parts of a flower.

Androecium Parts

Androgynophore. Stalk supporting the androecium and gynoecium in some flowers. Figure 1806.

Androphore. Stalk supporting a group of stamens.

Anther. The expanded, apical, pollen bearing portion of the stamen. Figure 1807.

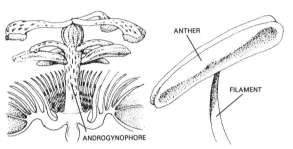

Figure 1806   **Figure 1807**

Anther sac. One of the pollen bearing chambers of the anther. Figure 1808.

Cell. A hollow cavity or compartment within a structure, as the cavity of the anther containing pollen. Figure 1808.

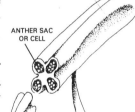

Figure 1808

Column. A structure formed by the union of staminal filaments. Figure 1809; the united filaments and style in the Orchidaceae. Figure 1810.

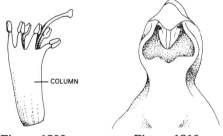

Figure 1809   **Figure 1810**

Connective. The portion of the stamen connecting the two pollen sacs of an anther. Figure 1811.

Filament. The stalk of the stamen which supports the anther. Figure 1807.

Gynandrium. A column bearing stamens and pistils. Figures 1810 and 1812.

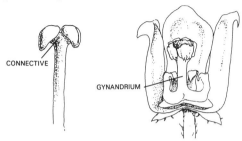

Figure 1811   **Figure 1812**

Phalange. Two or more stamens joined by their filaments. Figure 1813.

Phalanx. See **Phalange**.

Pollen. The mature microspore or developing male gametophyte of a seed plant, produced in the microsporangium of a gymnosperm or in the anther of an angiosperm. Figure 1814.

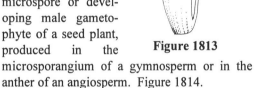

Figure 1813

Pollinium (pl. **pollinia**). A mass of waxy pollen grains transported as a unit in many members of the Orchidaceae and Asclepiadaceae. Figure 1815.

Figure 1814   **Figure 1815**

Sac. A bag-shaped compartment, as the cavity of an anther. See **anther sac**.

Stalk. See **filament**.

Stamen (pl. **stamens, stamina**). The male reproductive organ of a flower, consisting of an anther and filament. Figure 1807.

Staminode (pl. **staminodia**). A modified stamen which is sterile, producing no pollen. Figure 1816.

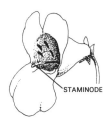

Figure 1816

Staminodium. See **staminode**.

Suture. The line of dehiscence of an anther. Figure 1817.

Theca (pl. **thecae**). A pollen sac or cell of the anther. See **anther sac**.

Translator. The connecting structure between the pollinia of adjacent anthers in the Asclepiadaceae. Figure 1818.

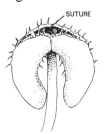

Figure 1817

Figure 1818

Stamen Types

Abortive. Not fully or properly developed; rudimentary. Figure 1819.

Fertile. Capable of bearing seeds; capable of bearing pollen.

Filantherous. Of a stamen with a distinct anther and filament. Figure 1820.

Figure 1819

Infertile. Sterile or inviable. See **abortive**.

Petalantherous. Of a stamen with a petaloid filament. See **petaloid**.

Petaloid. Petal-like in appearance. Figure 1821.

Polliniferous. Bearing pollen.

Rudimentary. Imperfectly developed; vestigial. See **abortive**.

Sterile. Infertile, as a stamen that does not bear pollen. See **abortive**.

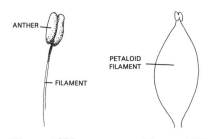

Figure 1820   **Figure 1821**

Stamen Number

Anandrous. Without stamens; lacking an androecium. Figure 1822.

Astemonous. Without stamens. Figure 1822.

Diandrous. With two stamens.

Haplostemonous. With as many stamens as petals. Figure 1823.

Figure 1822

Monandrous. With a single stamen.

Octandrous. With eight stamens.

Octostemonous. With eight stamens.

Oligandrous. With few stamens.

Pentandrous. With five stamens.

Polyandrous. With many stamens (usually more than ten). Figure 1824.

Polystemonous. With many stamens. Figure 1824.

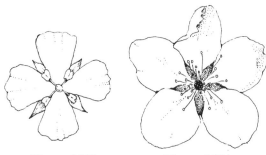

Figure 1823   **Figure 1824**

Stamen Arrangement

Alternate. Stamens borne between the petals. Figure 1825. (compare **opposite**)

Antepetalous. Directly in front of (opposite) the petals. Figure 1826.

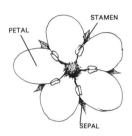

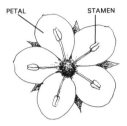

Figure 1825   **Figure 1826**

Antesepalous. Directly in front of (opposite) the sepals. Figure 1825.

Antipetalous. See **Antepetalous**.

Antisepalous. See **Antesepalous**.

Didynamous. With two pairs of stamens of unequal length; occurring in pairs. Figure 1827.

Figure 1827

Diplostemonous. With two series of stamens, the outer series opposite the sepals and the inner series opposite the petals; with twice as many stamens as petals. Figure 1828.

Excurved. Curving outward, away from the axis. Figure 1829.

Figure 1828   **Figure 1829**

Exserted. Stamens protruding from the corolla. Figure 1830.

Extrorse. Turned outward, away from the axis. Figure 1829. (compare **introrse**)

Haplostemonous. With one series of stamens. Figure 1831.

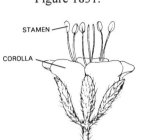

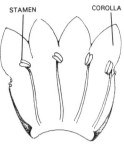

Figure 1830   **Figure 1831**

Included. Stamens not projecting beyond the corolla. Figure 1832.

Incurved. Curving inward, toward the axis. Figure 1833.

Introrse. Turned inward, toward the axis. Figure 1833. (compare **extrorse**)

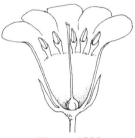

Figure 1832   **Figure 1833**

Obdiplostemonous. Having two whorls of stamens, the outer whorl opposite the petals and the inner whorl opposite the sepals. Figure 1834.

Opposite. Stamens borne on the same radius as the petals. Figure 1826. (compare **alternate**)

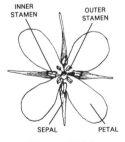

Figure 1834

Phaenantherous. With stamens exserted from the corolla. Figure 1830.

Polyadelphous. Borne in several distinct groups. Figure 1835. (compare **monadelphous** and **diadelphous**)

Tetradynamous. Having four long and two short stamens, as in most of the Cruciferae (Brassicaceae). Figure 1836.

Figure 1835   **Figure 1836**

Tridynamous. With stamens arranged in two groups of three.

Stamen Fusion

Adherent. Sticking together of unlike parts, as the anthers to the style. The attachment is not as firm or solid as **adnate**.

Adnate. Fusion of unlike parts, as the stamens to the corolla. Figure 1837. (compare **connate**)

Apostemonous. With separate stamens. Figure 1838.

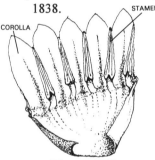

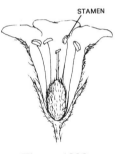

Figure 1837   **Figure 1838**

Appressed. Pressed close or flat against another organ.

Approximate. Borne close together, but not fused.

Coalescent. United together to form a single unit. Figure 1839.

Coherent. Sticking together of like parts. The attachment is not as firm or solid as **connate**.

Connate. Fusion of like parts, as the fusion of staminal filaments into a tube. Figure 1840. (compare **adnate**)

Figure 1839

Connivent. Converging, but not actually fused or united.

Diadelphous. Stamens united into two, often unequal, sets by their filaments. Figure 1841.

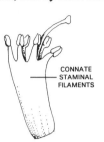

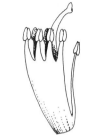

Figure 1840   **Figure 1841**

Distinct. Separate; not attached to like parts. Figure 1838. (compare **connate**)

Epipetalous. Attached to the petals. Figure 1842.

Free. Not attached to other organs. Figure 1838.

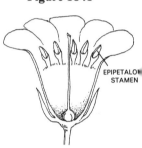

Figure 1842

Gynandrous. With the stamens adnate to the pistil. Figure 1843.

Gynostemial. See **gynandrous**.

Monadelphous. Stamens united by the filaments and forming a tube around the gynoecium. Figure 1844.

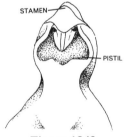

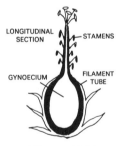

Figure 1843   **Figure 1844**

Petalostemonous. With the staminal filaments fused to the corolla and the anthers free. Figure 1845.

Polyadelphous. Borne in several distinct groups, as the stamens of some flowers. Figure 1846.

Synandrous. With united anthers. Figure 1839.

Syngenesious. With stamens united by their anthers. Figure 1839.

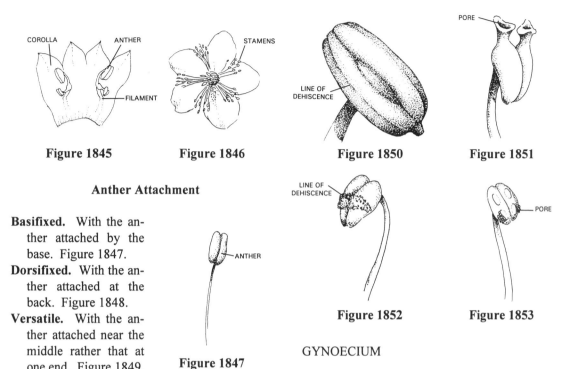

Figure 1845   **Figure 1846**   **Figure 1850**   **Figure 1851**

Anther Attachment

Basifixed. With the anther attached by the base. Figure 1847.

Dorsifixed. With the anther attached at the back. Figure 1848.

Versatile. With the anther attached near the middle rather that at one end. Figure 1849.

Figure 1847

Figure 1852   **Figure 1853**

GYNOECIUM

The female reproductive parts of a flower.

Gynoecium Parts

Androgynophore. Stalk supporting the androecium and gynoecium in some flowers. Figure 1854.

Carpel. A simple pistil formed from one modified leaf, or that part of a compound pistil formed from one modified leaf. Figure 1855.

Figure 1848   **Figure 1849**

Anther Dehiscence

Longitudinal. Dehiscing along the long axis of the anther. Figure 1850.

Poricidal. Dehiscing through a pore. Figure 1851.

Transverse. Dehiscing at a right angle to the long axis of the anther. Figure 1852.

Valvular. Dehiscing through flap-covered pores. Figure 1853.

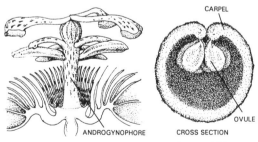

Figure 1854   **Figure 1855**

Carpopodium. A stipe supporting an ovary. Figure 1856.

Cell. A hollow cavity or compartment within an ovary; a locule. Figure 1857.

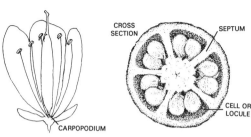

Figure 1856

Figure 1857

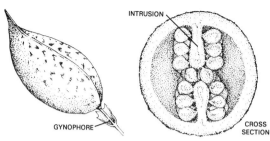

Figure 1861

Figure 1862

Chalaza. The part of an ovule or seed where the integuments are connected to the nucellus, at the opposite end from the micropyle. Figure 1858.

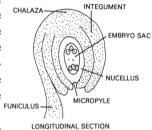

LONGITUDINAL SECTION

Figure 1858

Dissepiment. See **septum**.

Embryo sac. The megagametophyte within the ovule of a flowering plant. Figure 1858.

Funicle. See **funiculus**.

Funiculus (pl. **funiculi**). The stalk connecting the ovule to the placenta; the stalk of a seed. Figure 1858.

Gynandrium. A column bearing stamens and pistils. Figure 1859.

Gynobase. An elongation or enlargement of the receptacle, as in the flowers of the Boraginaceae. Figure 1860.

Figure 1859

Figure 1860

Gynophore. An elongated stalk bearing the pistil in some flowers. Figure 1861.

Integument. The covering of the ovule which will become the seed coat. Figure 1858.

Intrusion. Protrusion into, as placentae into the cell of an ovary. Figure 1862.

Kernel. See **nucellus**.

Locule. The chamber or cavity of an ovary containing the ovules. Figure 1857.

Loculus (pl. **loculi**). See **locule**.

Micropyle. The opening in the integuments of the ovule. Figure 1858.

Nucellus. The part of the ovule just beneath the integuments and surrounding the female gametophyte. Figure 1858.

Ovary. The expanded basal portion of the pistil that contains the ovules. Figure 1863.

Ovule. An immature seed; the megasporangium and surrounding integuments of a seed plant. Figure 1858.

Pistil. The female reproductive organ of a flower, typically consisting of a stigma, style, and ovary. Figure 1863.

Placenta (pl. **placentae**). The portion of the ovary bearing ovules. Figure 1864.

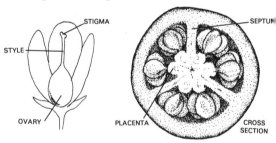

Figure 1863

Figure 1864

Podogyne. See **carpopodium**.

Rostellum. A small beak; an extension from the upper edge of the stigma in orchids. Figure 1865.

Septum (pl. **septa**). A partition, as the partitions separating the locules of an ovary. Figures 1857 and 1864.

Stigma. The portion of the pistil which is receptive to pollen. Figure 1863.

Stipe. A stalk supporting a structure, as the stalk attaching the ovary to the receptacle in some flowers. Figure 1866.

Stylocarpepodic. With a style and a stipe. Figure 1872.

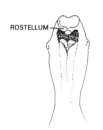

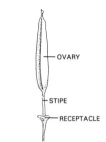

Figure 1865   **Figure 1866**

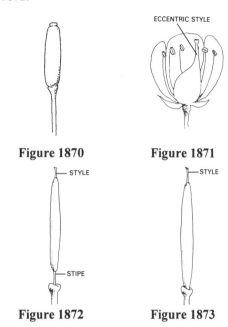

Figure 1870   **Figure 1871**

Style. The usually narrowed portion of the pistil connecting the stigma to the ovary. Figure 1863.

Stylopodium. A disk-like expansion or enlargement at the base of the style in the Umbelliferae (Apiaceae). Figure 1867.

Figure 1867

Figure 1872   **Figure 1873**

Carpel Number

Acarpous. Without carpels. Figure 1874.

Bicarpellate. With two carpels. Figure 1875.

Carpel Types

Astylocarpellous. Lacking a style and a stipe. Figure 1868.

Astylocarpepodic. Without a style, but with a stipe. Figure 1869.

Figure 1874   **Figure 1875**

Figure 1868   **Figure 1869**

Astylous. Without a style. Figure 1870.

Eccentric. Off-center; not positioned directly on the central axis. Figure 1871.

Stipitate. Borne on a stipe or stalk. Figure 1872.

Stylocarpellous. With a style, but without a stipe. Figure 1873.

Dicarpellate. See **bicarpellate**.

Digynous. With two pistils. Figure 1876.

Monocarpous. With one carpel. Figure 1877.

Monogynous. See **monocarpous**.

Octogynous. With eight pistils or styles. Figure

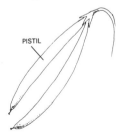

Figure 1876

1878.

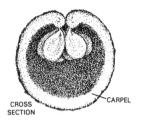

Figure 1877

Figure 1878

Polycarpous. With many carpels. Figure 1879.

Polygynous. With many pistils or styles. Figure 1879.

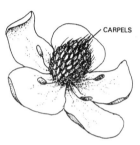

Figure 1879

Pseudomonomerous. A structure which appears to be simple, though actually derived from the fusion of separate structures, as a pistil which appears to be composed of a single carpel, though actually composed of two or more carpels.

Stylodious. See **unicarpellous**.

Tricarpellary. With three carpels. Figure 1880.

Unicarpellous. With a single, free carpel. Figure 1881.

Figure 1880

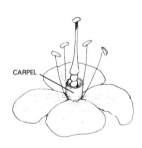

Figure 1881

Carpel Fusion

Apocarpous. A flower with carpels forming separate pistils, as in a buttercup. Figure 1882. (compare **syncarpous**)

Compound ovary. An ovary of two or more carpels. Figure 1880.

Free. Not attached to other organs. Figure 1883.

Gynandrial. See **gynandrous**.

Figure 1882

Figure 1883

Gynandrous. With the stamens adnate to the pistil. Figure 1884.

Gynostemial. See **gynandrous**.

Semicarpous. With ovaries of carpels partly fused, the styles and stigmas separate. Figure 1885.

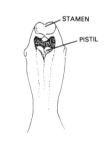

Figure 1884

Simple ovary. An ovary composed of only one carpel. Figure 1881.

Syncarpous. With united carpels. Figure 1886. (compare **apocarpous**)

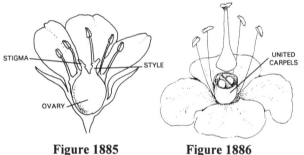

Figure 1885   Figure 1886

Ovary Position

Half-inferior. Attached below the lower half, as a flower with a hypanthium that is fused to the lower half of the ovary, giving the appearance that the other floral whorls are arising from about the middle of the ovary. Figure 1887.

Inferior. Attached beneath, as an ovary that is attached beneath the point of attachment of the other floral whorls which appear, therefore, to arise from the top of the ovary. Figure 1888.

Superior. Attached above, as an ovary that is attached above the point of attachment of the other floral whorls. Figure 1889.

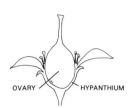

Figure 1887

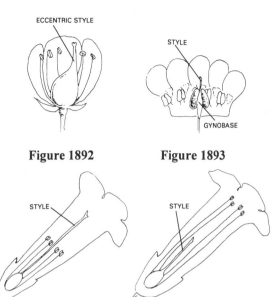

Figure 1892   **Figure 1893**

Figure 1894   **Figure 1895**

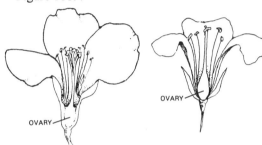

Figure 1888   **Figure 1889**

Style Forms

Astylous. Without a style. Figure 1890.

Bifid. Deeply two-cleft or two-lobed, usually from the tip. Figure 1891.

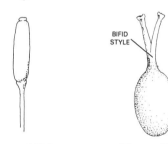

Figure 1890   **Figure 1891**

Eccentric. Off-center; not positioned directly on the central axis. Figure 1892.

Gynobasic style. A style which is attached to the gynobase as well as to the carpels. Figure 1893.

Heterostylic. With styles of different lengths in flowers of the same species. Figures 1894 and 1895.

Heterostylous. See **heterostylic**.

Homostylic. With styles of more or less constant length in flowers of the same species.

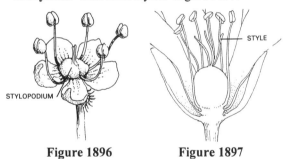

Homostylous. See **homostylic**.

Macrostylous. With a long style. Figure 1894.

Monostylous. With a single style. Figure 1894.

Stylopodic. With a stylopodium. Figure 1896.

Tristylous. With three styles. Figure 1897.

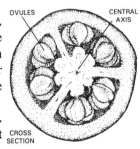

Figure 1896   **Figure 1897**

Placentation

Axile placentation. Ovules attached to the central axis of an ovary with two or more locules. Figure 1898.

Basal placentation. Ovules positioned at the base of a single-

Figure 1898

loculed ovary. Figure 1899.

Free-central placentation. Ovules attached to a free-standing column in the center of a unilocular ovary. Figure 1900.

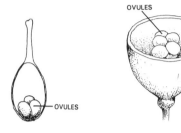

Figure 1899   **Figure 1900**

Marginal placentation. Ovules attached to the juxtaposed margins of a simple pistil. Figure 1901.

Parietal placentation. Ovules attached to the walls of the ovary. Figure 1902.

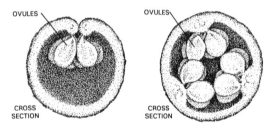

Figure 1901   **Figure 1902**

Ovule Types

Amphitropous ovule. An ovule which is half-inverted and straight, with the hilum lateral. Figure 1903.

Anatropous ovule. An ovule which is inverted and straight with the micropyle situated next to the funiculus. Figure 1904.

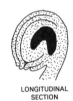

Figure 1903

Campylotropous ovule. An ovule which is curved so that the micropyle is positioned near the funiculus and the chalaza. Figure 1905.

Hemianatropous ovule. See **hemitropous ovule**.

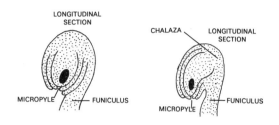

Figure 1904   **Figure 1905**

Hemitropous ovule. An ovule which is half-inverted so that the funiculus is attached near the middle with the micropyle at a right angle. Figure 1906.

Orthotropous ovule. An ovule which is straight and erect. Figure 1907.

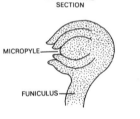

Figure 1906   **Figure 1907**

FRUITS

A ripened ovary and any other structures which are attached and ripen with it.

FRUIT PARTS

Article. Section of a fruit separated from others by a constricted joint. Figure 1908.

Carpophore. A slender prolongation of the receptacle between the carpels as a central axis, as in the fruits of some members of the Umbelliferae (Apiaceae) and the Geraniaceae. Figure 1909.

Cell. A hollow cavity or compartment within an ovary containing

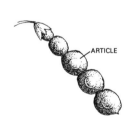

Figure 1908

ovules; a locule. Figure 1910.

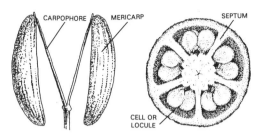

Figure 1909   **Figure 1910**

Commissure. The face by which two carpels join one another, as in the Umbelliferae (Apiaceae). Figure 1911.

Cupule. A cup-shaped involucre, as in an acorn. Figure 1912.

Figure 1911   **Figure 1912**

Dissepiment. Same as **septum**.

Endocarp. The inner layer of the pericarp of a fruit. Figure 1913. (compare **mesocarp** and **exocarp**)

Epicarp. Same as **exocarp**.

Exocarp. The outer layer of the pericarp of a fruit. Figure 1913. (compare **mesocarp** and **endocarp**)

Funiculus (pl. **funiculi**). The stalk of a seed. Figure 1914.

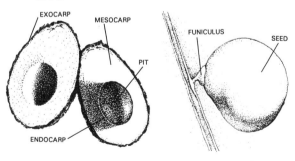

Figure 1913   **Figure 1914**

Gynophore. An elongated stalk bearing the pistil in some flowers.

Hypanthium. A cup-shaped extension of the floral axis usually formed from the union of the basal parts of the calyx, corolla, and androecium, commonly surrounding or enclosing the pistils. Figure 1915.

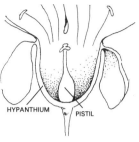

Figure 1915

Locule. The chamber or cavity of an ovary containing the seed. Figure 1910.

Loculus (pl. **loculi**). See **locule**.

Mericarp. A section of a schizocarp; one of the two halves of the fruit in the Umbelliferae (Apiaceae). Figure 1909.

Mesocarp. The middle layer of the pericarp of a fruit. Figure 1913. (compare **endocarp** and **exocarp**)

Operculum. A small lid, such as the deciduous cap of a circumscissile capsule. Figure 1916.

Ovary. The expanded basal portion of the pistil that contains the ovules; the immature fruit. Figure 1917.

Figure 1916   **Figure 1917**

Pericarp. The wall of the fruit. Figure 1918.

Pit. The stony endocarp of a drupe, as in a peach or cherry. Figure 1913.

Replum. Partition or septum between the two valves or compartments of silicles or

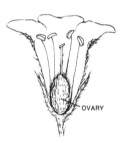

Figure 1918

siliques in the Cruciferae (Brassicaceae). Figure 1919.

Seed. A ripened ovule. Figure 1914.

Segment. A section of a fruit. Figure 1920.

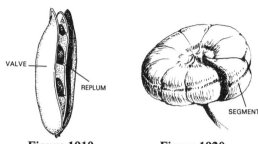

Figure 1919   **Figure 1920**

Septum (pl. **septa**). A partition, as the partitions separating the locules of an ovary. Figure 1910.

Stipe. A stalk attaching the fruit to the receptacle. Figure 1921.

Stone. The hard, woody endocarp enclosing the seed of a drupe. Figure 1922.

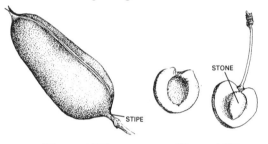

Figure 1921   **Figure 1922**

Stylopodium. A disklike expansion or enlargement at the base of the style in the Umbelliferae (Apiaceae). Figure 1923.

Suture. A line of fusion; the line of dehiscence of a fruit. Figure 1924.

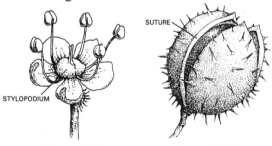

Figure 1923   **Figure 1924**

Valve. One of the segments of a dehiscent fruit, separating from other such segments at maturity. Figure 1925.

Vitta (pl. **vittae**). An oil tube in the carpel walls of the fruits of the Umbelliferae (Apiaceae). Figure

1926.

Figure 1925   **Figure 1926**

FRUIT TYPES (Figure 1927.)

Accessory fruit. A fleshy fruit developing from a succulent receptacle rather than the pistil. The ripened ovaries are small achenes on the surface of the receptacle, as in the strawberry.

Achene. A small, dry, indehiscent fruit with a single locule and a single seed, and with the seed attached to the ovary wall at a single point, as in the sunflower.

Acorn. The hard, dry, indehiscent fruit of oaks, with a single, large seed and a cuplike base.

Aggregate fruit. Usually applied to a cluster or group of small fleshy fruits originating from a number of separate pistils in a single flower, as in the clustered drupelets of the raspberry.

Akene. See **achene**.

Anthocarp. A fruit with some portion of the flower besides the pericarp persisting.

Aril. A fleshy thickening of the seed coat which resembles a true fruit, as in *Taxus*.

Berry. A fleshy fruit developing from a single pistil, with several or many seeds, as the tomato. Sometimes applied to any fruit which is fleshy or pulpy throughout, i.e. lacking a pit or core.

Bur. A fruit armed with often hooked or barbed spines or appendages.

Capsule. A dry, dehiscent fruit composed of more than one carpel.

Cariopsis. See **caryopsis**.

Caryopsis. A dry, one-seeded, indehiscent fruit with the seed coat fused to the pericarp, as in the fruits of the grass family; a grain.

Circumscissile capsule. A capsule dehiscing along a transverse circular line, so that the top separates

Key to Common Fruit Types

1 Fruit formed from more than one flower.
  2 Fruit consisting primarily of receptacle tissue, the ripened ovaries borne inside of the hollow, inverted receptacle. . . . . **Syconium**
  2 Fruit consisting of many tightly clustered ripened ovaries. . . . . **Multiple**
1 Fruit formed from a single flower.
  3 Fruit of more than one ovary.
    4 Carpels enclosed, borne on the wall of a globose hypanthium. . . . . **Hip**
    4 Carpels not enclosed, not borne on the wall of a hypanthium.
      5 Pistils developing into fleshy drupelets on a non-fleshy receptacle. . . . . **Aggregate**
      5 Pistils developing into achenes on a fleshy receptacle. . . . . **Accessory**
  3 Fruit of a single ovary.
    6 Fruit dry at maturity.
      7 Fruit dehiscent at maturity.
        8 Fruit composed of more than one carpel.
          9 Carpels two, separated by a persistent, translucent septum.
            10 Fruit less than twice longer than wide. . . . . **Silicle**
            10 Fruit more than twice longer than wide. . . . . **Silique**
          9 Carpels two or more, not separated by a persistent, translucent septum. **(Capsule)**
            11 Capsule opening along a transverse circular line, the top separating like a lid.
              . . . . **Circumscissile capsule**
            11 Capsule opening along longitudinal lines or by pores.
              12 Capsule opening by pores. . . . . **Poricidal capsule**
              12 Capsule opening along longitudinal lines.
                13 Capsule dehiscing through the locules. . . . . **Loculicidal capsule**
                13 Capsule dehiscing through the septae. . . . . **Septicidal capsule**
        8 Fruit composed of a single carpel.
          14 Fruit opening along a single line of dehiscence. . . . . **Follicle**
          14 Fruit opening along two lines of dehiscence.
            15 Fruit not obviously constricted between the seeds. . . . . **Legume**
            15 Fruit obviously constricted between the seeds. . . . . **Loment**
      7 Fruit indehiscent at maturity.
        16 Fruit splitting at maturity, but carpels not dehiscing to release seeds. . . . . **Schizocarp**
        16 Fruit not splitting at maturity.
          17 Fruit winged. . . . . **Samara**
          17 Fruit not winged.
            18 Seed inseparably fused to the ovary wall. . . . . **Caryopsis, Grain**
            18 Seed not inseparably fused to the ovary wall.
              19 Fruit wall bladdery-inflated. . . . . **Utricle**
              19 Fruit wall not bladdery-inflated.
                20 Fruit wall hard and tough.
                  21 Fruit very small. . . . . **Nutlet**
                  21 Fruit larger. . . . . **Nut**
                20 Fruit wall not particularly hard and tough. . . . . **Achene**

```
6 Fruit fleshy at maturity.
   22 Seed one.
      23 Fruit very small. . . . . . . . . . . . . . . . . . . . . . . . . . . . . . . . . . . . . . . . . . . . . . . Drupelet
      23 Fruit larger. . . . . . . . . . . . . . . . . . . . . . . . . . . . . . . . . . . . . . . . . . . . . . . . . Drupe
   22 Seeds more than one.
      24 Fruit surrounded by the fleshy receptacle. . . . . . . . . . . . . . . . . . . . . . . . . . . . . . . Pome
      24 Fruit not surrounded by the receptacle.
         25 Fruit with a tough rind. . . . . . . . . . . . . . . . . . . . . . . . . . . . . . . . . . . . . . . Pepo
         25 Fruit lacking a tough rind. . . . . . . . . . . . . . . . . . . . . . . . . . . . . . . . . . . . Berry
```

like a lid.

Cremocarp. See **schizocarp**.

Drupe. A fleshy, indehiscent fruit with a stony endocarp surrounding a usually single seed, as in a peach or cherry.

Drupelet. A small drupe, as in the individual segments of a raspberry fruit.

Follicle. A dry, dehiscent fruit composed of a single carpel and opening along a single side, as a milkweed pod.

Grain. A seed-like structure, as in the fruit of some *Rumex* species; a caryopsis.

Hesperidium. A fleshy berry-like fruit with a tough rind, as a lemon or orange.

Hip. A berry-like structure composed of an enlarged hypanthium surrounding numerous achenes.

Legume. A dry, dehiscent fruit derived from a single carpel and usually opening along two lines of dehiscence, as a pea pod.

Loculicidal capsule. A capsule dehiscing through the locules of a fruit rather than through the septa. (compare **septicidal** and **poricidal**)

Loment. A legume which is constricted between the seeds.

Lomentum (pl. **lomenta**). See **loment**.

Multiple fruit. A fruit formed from several separate flowers crowded on a single axis, as a mulberry or pineapple.

Nut. A hard, dry, indehiscent fruit, usually with a single seed.

Nutlet. A small nut; one of the lobes or sections of the mature ovary of some members of the Boraginaceae, Verbenaceae, and Labiatae (Lamiaceae).

Pepo. A fleshy, indehiscent, many-seeded fruit with a tough rind, as a melon or a cucumber.

Pod. Any dry, dehiscent fruit, especially a legume or follicle.

Pome. A fleshy, indehiscent fruit derived from an inferior, compound ovary, consisting of a modified floral tube surrounding a core, as in an apple.

Poricidal capsule. A capsule opening by pores, as in a poppy.

Pseudocarp. A fruit which develops from the receptacle rather than the ovary, as in a pome.

Pyxidium. See **pyxis**.

Pyxis. A circumscissile capsule, the top coming off as a lid.

Samara. A dry, indehiscent, winged fruit.

Schizocarp. A dry, indehiscent fruit which splits into separate one-seeded segments (carpels) at maturity.

Septicidal capsule. A capsule dehiscing through the septa and between the locules. (compare **loculicidal** and **poricidal**)

Silicle. A dry, dehiscent fruit of the Cruciferae (Brassicaceae), typically less than twice as long as wide, with two valves separating from the persistent placentae and septum (replum).

Silique. A dry, dehiscent fruit of the Cruciferae (Brassicaceae), typically more than twice as long as wide, with two valves separating from the persistent placentae and septum (replum).

Syconium. The fruit of a fig, consisting of an entire ripened inflorescence with a hollow, inverted receptacle bearing flowers internally.

Syncarp. A multiple fruit.

Utricle. A small, thin-walled, one-seeded, more or less bladdery-inflated fruit.

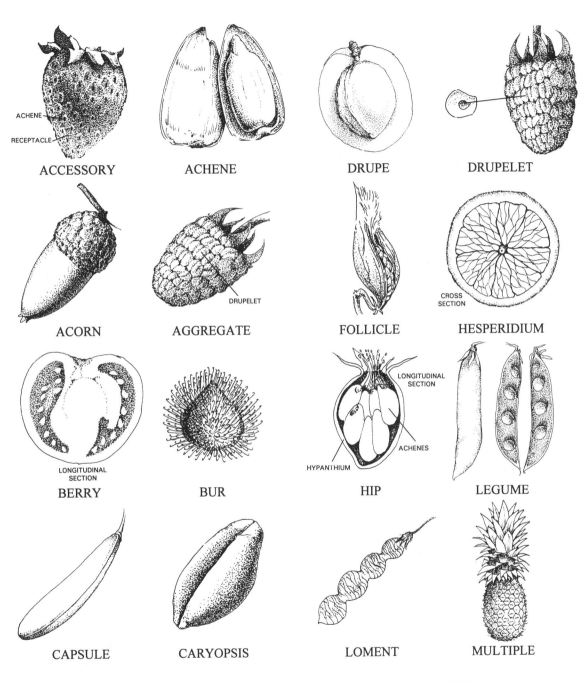

Figure 1927a.   **Figure 1927b.**

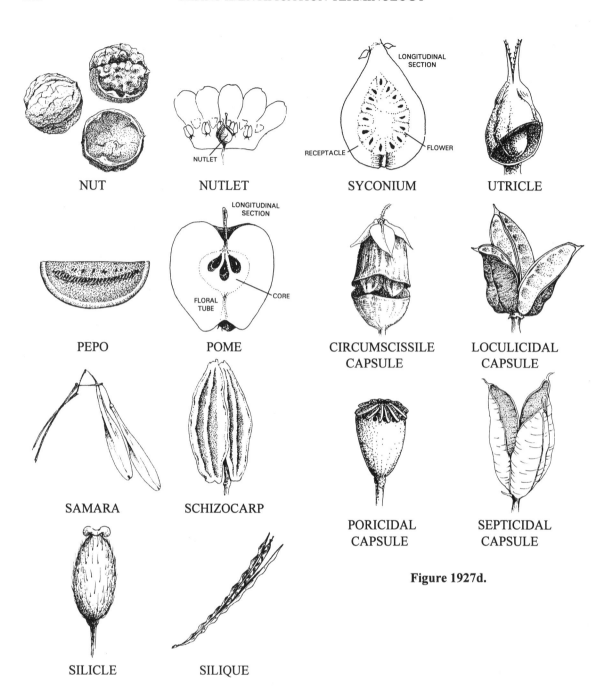

NUT

NUTLET

SYCONIUM

UTRICLE

PEPO

POME

CIRCUMSCISSILE CAPSULE

LOCULICIDAL CAPSULE

SAMARA

SCHIZOCARP

PORICIDAL CAPSULE

SEPTICIDAL CAPSULE

Figure 1927d.

SILICLE

SILIQUE

Figure 1927c.